BEARINGS

BEARINGS

Edited by

M. J. NEALE
OBE, BSc (Eng), DIC, FCGI, WhSch, FEng, FIMechE

A TRIBOLOGY HANDBOOK

Butterworth-Heinemann Ltd
Linacre House, Jordan Hill, Oxford OX2 8DP

$\mathcal{R}$ A member of the Reed Elsevier plc group

OXFORD LONDON BOSTON
NEW DEHLI SINGAPORE SYDNEY
TOKYO TORONTO WELLINGTON

First published 1993
Reprinted 1995

British Library Cataloguing in Publication Data

Bearings. – 2 Rev. ed of ''Tribology
   Handbook'' – (Tribology Handbook)
   I. Neale, M. J. II. Series
   621.8

ISBN 0 7506 0979 6

Printed and bound in Great Britain by
Athenæum Press Ltd, Gateshead, Tyne & Wear

Contents

Editor's Preface

This handbook gives practical guidance on bearings in a form intended to provide easy and rapid reference. It is based on material published in the first edition of the *Tribology Handbook* and has been updated and matched to international requirements.

Each section has been written by an author who is expert in the field and who in addition to understanding the related basic principles, also has extensive practical experience in his subject area.

The individual contributors are listed and the editor gratefully acknowledges their assistance and that of all other people who have helped him in the checking and compilation of this revised volume.

Michael Neale
Neale Consulting Engineers Ltd
Farnham 1992

Contributors

Section	Author
Selection of bearing type and form	M. J. Neale OBE, BSc(Eng), DIC, FCGI, WhSch, FEng, FIMechE
Selection of journal bearings	M. J. Neale OBE, BSc(Eng), DIC, FCGI, WhSch, FEng, FIMechE
Selection of thrust bearings	P. B. Neal BEng, PhD, CEng, MIMechE
Plain bearing materials	P. T. Holligan BSc(Tech), FIM, J. M. Conway Jones BSc, PhD, DIC, ACGI
Dry rubbing bearings	J. K. Lancaster PhD, DSc, FInstP
Porous metal bearings	V. T. Morgan AIM, MIMechE
Grease, wick and drip fed journal bearings	W. H. Wilson BSc(Eng), CEng, MIMechE
Ring and disc fed journal bearings	F. A. Martin CEng, FIMechE
Steady load pressure fed journal bearings	F. A. Martin CEng, FIMechE
High speed bearings and rotor dynamics	M. J. Neale OBE, BSc(Eng), DIC, FCGI, WhSch, FEng, FIMechE
Crankshaft bearings	D. de Geurin CEng, FIMechE
Plain bearing form and installation	J. M. Conway Jones BSc, PhD, DIC, ACGI
Oscillatory journal bearings	K. Jakobsen LicTechn
Spherical bearings	D. Bastow BSc(Eng), CEng, FIMechE, MConsE, MSAE, MSIA(France)
Plain thrust bearings	P. B. Neal BEng, PhD, CEng, MIMechE
Profiled pad thrust bearings	P. B. Neal, BEng, PhD, CEng, MIMechE
Tilting pad thrust bearings	A. Hill CEng, FIMechE, FIMarE
Hydrostatic bearings	W. B. Rowe BSc, PhD, DSc, CEng, FIMechE, FIEE
Gas bearings	A. J. Munday BSc(Tech), CEng, MIMechE
Selection of rolling bearings	D. G. Hjertzen CEng, MIMechE
Rolling bearing materials	D. B. Jones CEng, MIMechE, P. L. Hurricks BSc, MSc
Rolling bearing installation	C. W. Foot CEng, MIMechE
Slide bearings	F. M. Stansfield BSc(Tech), CEng, MIMechE, A. E. Young BEng, CEng, MIMechE, AMCT
Instrument jewels	G. F. Tagg BSc, PhD, CEng, FInstP, FIEE, FIEEE
Flexures and knife edges	A. B. Crease MSc, ACGI, CEng, MIMechE
Electromagnetic bearings	G. Fletcher BSc, CEng, MIMechE
Bearing surface treatments and coatings	M. J. Neale OBE, BSc(Eng), DIC, FCGI, WhSch, FEng, FIMechE

Bearings allow relative movement between the components of machines, while providing some type of location between them.

The form of bearing which can be used is determined by the nature of the relative movement required and the type of constraints which have to be applied to it.

Relative movement between machine components and the constraints applied

Constraint applied to the movement	Continuous movement	Oscillating movement
About a point	The movement will be a rotation, and the arrangement can therefore make repeated use of accurate surfaces	If only an oscillatory movement is required, some additional arrangements can be used in which the geometric layout prevents continuous rotation
About a line	The movement will be a rotation, and the arrangement can therefore make repeated use of accurate surfaces	If only an oscillatory movement is required, some additional arrangements can be used in which the geometric layout prevents continuous rotation
Along a line	The movement will be a translation. Therefore one surface must be long and continuous, and to be economically attractive must be fairly cheap. The shorter, moving component must usually be supported on a fluid film or rolling contact for an acceptable wear rate	If the translational movement is a reciprocation, the arrangement can make repeated use of accurate surfaces and more mechanisms become economically attractive
In a plane	If the movement is a rotation, the arrangement can make repeated use of accurate surfaces	If the movement is rotational and oscillatory, some additional arrangements can be used in which the geometric layout prevents continuous rotation
	If the movement is a translation one surface must be large and continuous and to be economically attractive must be fairly cheap. The smaller moving component must usually be supported on a fluid film or rolling contact for an acceptable wear rate	If the movement is translational and oscillatory, the arrangement can make repeated use of accurate surfaces and more mechanisms become economically attractive

For both continuous and oscillating movement, there will be forms of bearing which allow movement only within a required constraint, and also forms of bearing which allow this movement among others.

The following tables give examples of both these forms of bearing, and in the case of those allowing additional movement, describe the effect which this can have on a machine design.

Selection of bearing type and form

Examples of forms of bearing suitable for continuous movement

Constraint applied to the movement	*Examples of arrangements which allow movement only within this constraint*	*Examples of arrangements which allow this movement but also have other degrees of freedom*	*Effect of the other degrees of freedom*
About a point	Gimbals	Ball on a recessed plate	Ball must be forced into contact with the plate
About a line	Journal bearing with double thrust location	Journal bearing	Simple journal bearing allows free axial movement as well
	Double conical bearing	Screw and nut	Gives some related axial movement as well
		Ball joint or spherical roller bearing	Allows some angular freedom to the line of rotation
Along a line	Crane wheel restrained between two rails	Railway or crane wheel on a track	These arrangements need to be loaded into contact. This is usually done by gravity. Wheels on a single rail or cable need restraint to prevent rotation about the track member
		Pulley wheel on a cable	
		Hovercraft or hoverpad on a track	
In a plane (rotation)	Double thrust bearing	Single thrust bearing	Single thrust bearing must be loaded into contact
In a plane (translation)		Hovercraft or hoverpad	Needs to be loaded into contact usually by gravity

Examples of forms of bearing suitable for oscillatory movement only

Constraint applied to the movement	*Examples of arrangements which allow movement only within this constraint*	*Examples of arrangements which allow this movement but also have other degrees of freedom*	*Effect of the other degrees of freedom*
About a point	Hookes joint	Cable connection between components	Cable needs to be kept in tension
About a line	Crossed strip flexure pivot	Torsion suspension	A single torsion suspension gives no lateral location
		Knife-edge pivot	Must be loaded into contact
		Rubber bush	Gives some axial and lateral flexibility as well
		Rocker pad	Gives some related translation as well. Must be loaded into contact
Along a line	Crosshead and guide bars	Piston and cylinder	Piston can rotate as well unless it is located by connecting rod
In a plane (rotation)		Rubber ring or disc	Gives some axial and lateral flexibility as well
In a plane (translation)	Plate between upper and lower guide blocks	Block sliding on a plate	Must be loaded into contact

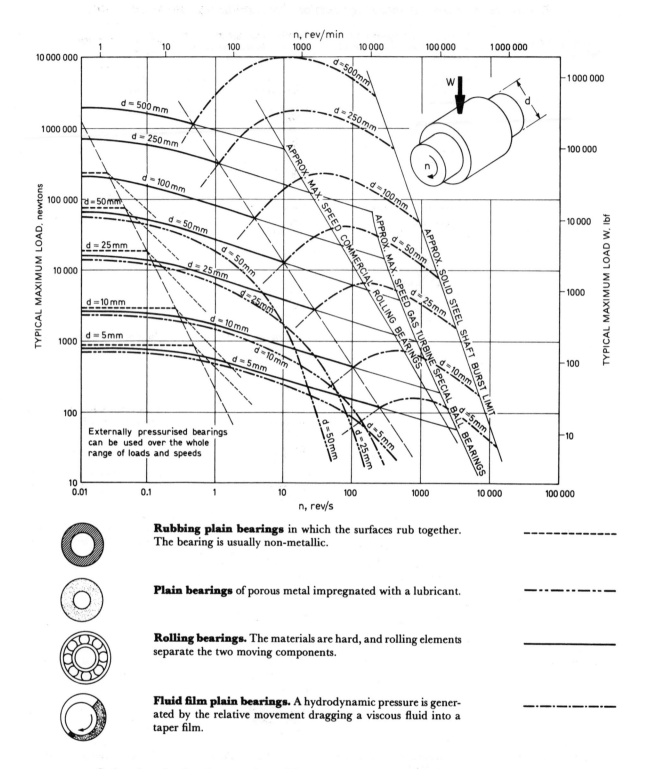

Rubbing plain bearings in which the surfaces rub together. The bearing is usually non-metallic.

Plain bearings of porous metal impregnated with a lubricant.

Rolling bearings. The materials are hard, and rolling elements separate the two moving components.

Fluid film plain bearings. A hydrodynamic pressure is generated by the relative movement dragging a viscous fluid into a taper film.

Selection by load capacity of bearings with continuous rotation

This figure gives guidance on the type of bearing which has the maximum load capacity at a given speed and shaft size. It is based on a life of 10 000 h for rubbing, rolling and porous metal bearings. Longer lives may be obtained at reduced loads and speeds. For the various plain bearings, the width is assumed to be equal to the diameter, and the lubricant is assumed to be a medium viscosity mineral oil.

In many cases the operating environment or various special performance requirements, other than load capacity, may be of overriding importance in the selection of an appropriate type of bearing. The tables give guidance for these cases.

Selection of journal bearings with continuous rotation for special environmental conditions

Type of bearing	High temp.	Low temp.	Vacuum	Wet and humid	Dirt and dust	External Vibration	Type of bearing
Rubbing plain bearings (non-metallic)	Good up to the temperature limit of material	Good	Excellent	Good but shaft must be incorrodible	Good but sealing helps	Good	
Porous metal plain bearings oil impregnated	Poor since lubricant oxidises	Fair; may have high starting torque	Possible with special lubricant	Good	Sealing essential	Good	
Rolling bearings	Consult makers above 150°C	Good	Fair with special lubricant	Fair with seals	Sealing essential	Fair; consult makers	
Fluid film plain bearings	Good to temperature limit of lubricant	Good; may have high starting torque	Possible with special lubricant	Good	Good with seals and filtration	Good	
Externally pressurised plain bearings	Excellent with gas lubrication	Good	No; lubricant feed affects vacuum	Good	Good; excellent when gas lubricated	Excellent	
General comments	Watch effect of thermal expansion on fits			Watch corrosion		Watch fretting	

Selection of journal bearings with continuous rotation for special performance requirements

Type of bearing	Accurate radial location	Axial load capacity as well	Low starting torque	Silent running	Standard parts available	Simple lubrication	Type of bearing
Rubbing plain bearings (non-metallic)	Poor	Some in most cases	Poor	Fair	Some	Excellent	
Porous metal plain bearings oil impregnated	Good	Some	Good	Excellent	Yes	Excellent	
Rolling bearings	Good	Yes in most cases	Very good	Usually satisfactory	Yes	Good when grease lubricated	
Fluid film plain bearings	Fair	No; separate thrust bearing needed	Good	Excellent	Some	Usually requires a circulation system	
Externally pressurised plain bearings	Excellent	No; separate thrust bearing needed	Excellent	Excellent	No	Poor; special system needed	

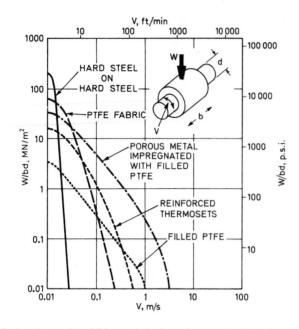

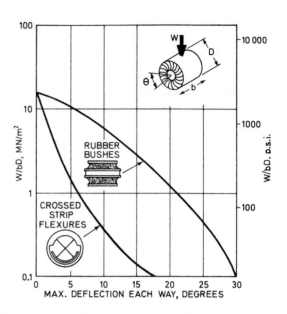

Selection of rubbing plain bearing materials for bushes with oscillatory movement, by maximum pressure and maximum value of average sliding speed. Rolling bearings in an equivalent arrangement usually can carry about 10 MN/m²

Selection of flexure bearings by external load pressure and the required deflection. If the centre of rotation does not have to be held constant, single strips or cables can be used

Selection of journal bearings with oscillating movement for special environments or performance

Type of bearing	Low friction	High temp.	Low temp.	Dirt and dust	External Vibration	Wet and humid	Type of bearing
Rubbing plain bearings	Good with PTFE	Good to the temp. limit of material	Very good	Good but sealing helps	Very good	Good but shaft must be incorrodible	
Porous metal plain bearings oil impregnated	Good	Poor since lubricant oxidises	Fair; friction can be high	Sealing is essential	Good	Good	
Rolling bearings	Very good	Consult makers above 150°C	Good	Sealing is essential	Poor	Good with seals	
Rubber bushes	Elastically stiff	Poor	Poor	Excellent	Excellent	Excellent	
Strip flexures	Excellent	Good	Very good	Excellent	Excellent	Good; watch corrosion	
Knife edge pivots	Very good	Good	Good	Good	Poor	Good; watch corrosion	

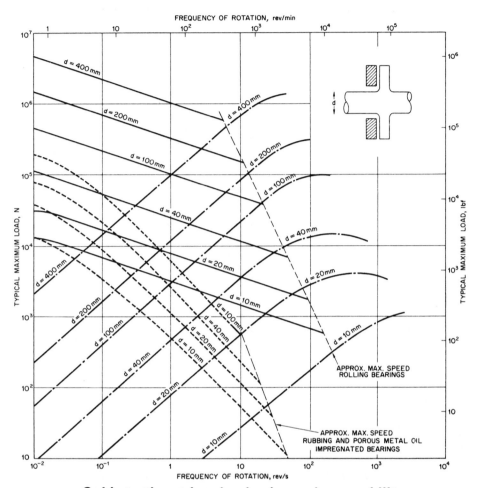

Guide to thrust bearing load-carrying capability

 Rubbing*‡ (generally intended to operate dry—life limited by allowable wear).

 Oil impregnated porous metal*‡ (life limited by lubricant degradation or dryout).

 Hydrodynamic oil film*† (film pressure generated by rotation—inoperative during starting and stopping).

 Rolling‡ (life limited by fatigue).

 Hydrostatic (applicable over whole range of load and speed—necessary supply pressure 3–5 times mean bearing pressure).

* Performance relates to thrust face diameter ratio of 2.
† Performance relates to mineral oil having viscosity grade in range 32–100 ISO 3448
‡ Performance relates to nominal life of 10 000 h.

This figure gives guidance on the maximum load capacity for different types of bearing for given speed and shaft size.

In many cases the operating environment or various special performance requirements, other than load capacity, may be of overriding importance in the selection of an appropriate type of bearing. The tables give guidance for these cases.

3 Selection of thrust bearings

Thrust bearing selection for special environmental conditions

Type of bearing	High temperature	Low temperature	Vacuum	Wet and humid	Dirt and dust	External vibration
Rubbing bearing (non-metallic)	Good—to temperature limit of material	Good	Excellent	Good with suitable shaft and runner material	Good—sealing helps	Good
Oil-impregnated porous metal bearing	Poor—lubricant oxidation	Fair—starting torque may be high	Possible with special lubricant	Good	Sealing necessary	Good
Rolling bearing	Above 100°C reduced load capacity Above 150°C consult makers	Good	Requires special lubricant	Fair with seals	Sealing necessary	Consult makers
Hydrodynamic film bearing	Good—to temperature limit of lubricant	Good—starting torque may be high	Possible with special lubricant	Good	Sealing necessary	Good
Hydrostatic film bearing	Good—to temperature limit of lubricant	Good	Not normally applicable	Good	Good—filtration necessary	Good
General comments	Consider thermal expansion and fits			Consider corrosion		Consider fretting

Thrust bearing selection for special performance requirements

Type of bearing	Accuracy of axial location	Low starting torque	Low running torque	Silent running	Suitability for oscillatory or intermittent movement	Availability of standard parts	Simplicity of lubrication system
Rubbing bearing (non-metallic)	Limited by wear	Poor	Poor	Fair	Yes	Some	Excellent
Oil-impregnated porous metal bearing	Good	Fair	Good	Good—until bearing dry-out	Yes	Good	Excellent
Rolling bearing	Good	Good	Good	Usually satisfactory	Yes	Excellent	Good—when grease lubricated
Hydrodynamic film bearing	Good	Fair	Good	Good	No	Some	Usually requires circulation system
Hydrostatic film bearing	Excellent	Excellent	Good	Good	Yes	No	Special system necessary

Requirements and characteristics of lubricated plain bearing materials

Physical property	Significance of property in service	Characteristics of widely used materials		
		White metals	Copper-base alloys	Aluminium-base alloys
Fatigue strength	To sustain imposed dynamic loadings at operating temperature	Adequate for many applications, but falls rapidly with rise of temperature	Wide range of strength available by selection of composition	Similar to copper-base alloys by appropriate selection of composition
Compressive strength	To support uni-directional loading without extrusion or dimensional change	As above	As above	As above
Embedd-ability	To tolerate and embed foreign matter in lubricant, so minimising journal wear	Excellent—unequalled by any other bearing materials	Inferior to white metals. Softer weaker alloys with low melting point constituent, e.g. lead; superior to harder stronger alloys in this category. These properties can be enhanced by provision of overlay, e.g. lead–tin or lead–indium, on bearing surface where appropriate	Inferior to white metals. Alloys with high content of low melting point constituent, e.g. tin or cadmium; superior in these properties to copper-base alloys of equivalent strength. Overlays may be provided in appropriate cases to enhance these properties
Conform-ability	To tolerate some mis-alignment or journal deflection under load			
Compati-bility	To tolerate momentary boundary lubrication or metal-to-metal contact without seizure			
Corrosion resistance	To resist attack by acidic oil oxidation products or water or coolant in lubricant	Tin-base white metals excellent in absence of sea-water. Lead-base white metals attacked by acidic products	Lead constituent, if present, susceptible to attack. Resistance enhanced by lead–tin or lead–tin–copper overlay	Good. No evidence of attack of aluminium-rich matrix even by alkaline high-additive oils

Physical properties, forms available, and applications of some white metal bearing alloys

Type of bearing	Physical properties			Forms available	Applications
	Melting range, °C	Hardness H_v at 20°C	Coefficient of expansion $\times 10^{-6}$/°C		
Tin-base white metal ISO 4381 Sn Sb8 Cu4 tin 89% antimony 7.5% copper 3.5%	239–312	~23–25	~23	Lining of thin-walled steel-backed half-bearings, split bushes and thrust washers; lining of bronze-backed components, unsplit bushes	Crankshaft bearings of ic engines and reciprocating compressors within fatigue range; FHP motor bushes; gas turbine bearings (cool end); camshaft bushes; general lubricated applications
Tin-base white metal ISO 4381 Sn Sb8 Cu4 Cd tin 87% antimony 8% copper 4% cadmium 1%	239–340	~27–32	~23	Lining of medium and thick-walled half-bearings and bushes; lining of direct-lined housings and connecting rods	Crankshaft and cross-head bearings of medium and large diesel engines within fatigue range; marine gearbox bearings; large plant and machinery bearings; turbine bearings
Lead-base white metal ISO 4381 Pb Sb10 Sn6 tin 6% antimony 10% copper 1% lead 83%	245–260	~26	~28	'Solid' die-castings; lining of steel, cast iron, and bronze components	General plant and machinery bearings operating at lower loads and temperatures

Note: for the sake of brevity the above table lists only two tin-base and one lead-base white metal. For other white metals and applications refer to ISO 4381 and bearing suppliers.

Physical properties, forms available, and applications of some copper-base alloy bearing materials

Type of bearing	Physical properties			Forms available	Applications
	Melting range, °C	Hardness H_v at 20°C	Coefficient of expansion $\times 10^{-6}/°C$		
Lead bronze ISO 4382/1 Cu Pb20 Sn5 copper 75% tin 5% lead 20%	Matrix ~900 Lead constituent ~327	45–70	~18	Machined cast components; as lining of steel-backed components	Machined bushes, thrust washers, slides, etc., for high-duty applications; as lining of thin and medium-walled heavily loaded IC engine crankshaft bearings, small-end and camshaft bushes, gearbox bushes; gas-turbine bearings, etc.
Lead bronze ISO 4382/1 Cu Pb10 Sn10 copper 80% tin 10% lead 10%	Matrix ~820 Lead constituent ~327	65–90	~18	As above. Hard, strong bronze	Machined bush and thrust washer applications; as lining of thin-walled split bushes for small-ends, camshafts, gearboxes, linkages, etc.
Lead bronze ISO 4382/1 Cu Pb9 Sn5 copper 85% tin 5% lead 9%	Matrix ~920 Lead constituent ~327	45–70	~18	Machined cast components, bars, tubes	Bushes, thrust washers, slides for wide range of applications
Phosphor bronze ISO 4382/1 Cu Sn10 P copper; remainder tin 10% min. phosphorus 0.5% min.	~800	70–150	~18	Machined cast components; bushes, bars, tubes, thrust washers, slides, etc.	Heavy load, high-temperature bush and slide applications, e.g. crankpress bushes, rolling mill bearings, gudgeon-pin bushes, etc.
Copper lead ISO 4383 Cu Pb30 copper 70% lead 30%	Matrix ~1050 Lead constituent ~327	35–45	Lining ~16	As lining of thin-, medium- or thick-walled half-bearings, bushes, and thrust washers	Crankshaft bearings for high- and medium-speed petrol and diesel engines; gas-turbine and turbo-charger bearings; compressor bearings; camshaft and rocker bushes. May be used with or without overlay
Lead bronze ISO 4383 Cu Pb24 Sn4 copper 74% lead 24% tin 4%	Matrix ~900 Lead constituent ~327	40–55	Lining ~18	As above	As above, for more heavily loaded applications, usually overlay plated for crank-shaft bearing applications

Note: for details of other lead bronzes, gunmetals, leaded gunmetals, and aluminium bronzes, consult ISO 4382/1, ISO 4383, EN 133 and bearing supplier.

Physical properties, forms available, and applications of some aluminium-base alloy bearing materials

Type of bearing	Physical properties			Forms available	Applications
	Melting range, °C	Hardness H_v at 20°C	Coefficient of expansion $\times 10^{-6}/°C$		
Duralumin-type material, no free low melting-point constituents	~550–650	~80–150 depending on composition and heat treatment	~22–24	Cast or wrought machined components	Bushes, slides, etc., for slow speed heavily loaded applications, e.g. small ends of medium and large diesel engines; general machinery bushes, etc.
Low tin aluminium alloy ISO 4383 Al Sn6 Cu tin 6% copper 1% nickel 1% aluminium remainder	Matrix ~650 Tin eutectic ~230	~45–60	~24	Cast or rolled machined components. As lining of steel-backed components. Usually overlay plated for crankshaft bearing applications	Unsplit bushes for small-ends, rockers, linkages, gearboxes. Crankshaft half-bearings for diesel engines and linings of thin- and medium-walled steel-backed crankshaft bearings for heavily loaded diesels and compressors; also as linings of steel-backed split bushes.
Aluminium silicon-cadmium alloy ISO 4383 Al Si4 Cd silicon 4% cadmium 1% aluminium remainder	Matrix ~650 Cadmium ~320	~55	~22	As lining of thin-walled half-bearings, split bushes and thrust washers. Usually overlay plated for crankshaft applications	Heavily loaded diesel engine crankshaft bearings; small-end, gearbox, rocker bushes, etc.
Aluminium tin silicon alloy ISO 4383 tin 10 or 12% silicon 4% copper 1 or 2% aluminium remainder	Matrix ~650 Tin eutectic ~230	55–65	~24	As lining of thin-walled half-bearings, split bushes and thrust washers. Used without an overlay	Heavily loaded crankshaft bearings for high-speed petrol and diesel engines. Used without an overlay
High tin aluminium alloy ISO 4383 Al Si20 Cu tin 20% copper 1% aluminium remainder	Matrix ~650 Tin eutectic ~230	~40	~24	As lining of thin- and medium-walled half-bearings, split bushes and thrust washers. Usually used without an overlay	Moderately loaded crankshaft bearings for high-speed petrol and diesel engines. Camshaft gearbox and linkage bushes; thrust washers

Overlay plating

Functions of an overlay

1 To provide bearing surface with good frictional properties, i.e. compatibility

2 To confer some degree of embeddability

3 To improve load distribution

4 To protect lead in lead-containing interlayer materials (e.g. copper–lead, lead bronze) from corrosion

Thickness of overlay

0.017 mm (0.0007 in) to 0.040 mm (0.0015 in) depending upon bearing loading and type and size of bearing

Typical overlay compositions

1 10–12% tin, remainder lead

2 10% tin, 2% copper, remainder lead
Tin and copper may be higher for increased corrosion resistance.
Where the maximum corrosion resistance is required with lead-tin-copper overlays or copper-lead a nickel interlayer 0.001 mm thick is used beneath the overlay

3 5–8% indium, remainder lead

4 20–40% tin, remainder aluminium applied by vapour deposition (sputter) on aluminium alloy substrates

4 Plain bearing materials

Relative load-carrying capacities of bearing materials, and recommended journal hardness

Material	Maximum dynamic loading		Recommended journal hardness
	MN/m^2	lbf/in^2	H_v min at 20°C
Tin and lead-base white metal linings ∼0.5 mm (0.020 in) thick	10.3–13.7	1500–2000	Soft journal (∼140) satisfactory
As above ∼0.1 mm (0.004 in) thick	>17.2	>2500	As above
70/30 copper–lead on steel	24–27.5	3500–4000	∼250
70/30 copper–lead on steel, overlay plated	27.5–31	4000–4500	∼230
Lead bronze, 20–25% lead, 3–5% tin, on steel	35–42	5000–6000	∼500
As above, overlay plated	42–52	6000–7500	∼230
Low lead (∼10%) lead bronzes, steel-backed	>48	>7000	∼500
Aluminium–6% tin, 'solid' or steel-backed	∼42	∼6000	∼500
As above, overlay plated	45–52	6500–7500	∼280
Aluminium–20% tin (reticular structure) on steel	∼42	∼6000	∼230
Aluminium–tin–silicon on steel	∼52	∼7500	∼250
Phosphor–bronze, chill or continuously cast	∼62	∼9000	∼500

Note: the above figures must be interpreted with caution, as they apply only to specific testing conditions. They should not be used for design purposes without first consulting the appropriate bearing supplier.

Fatigue strength and relative compatibility of some bearing alloys (courtesy: Glacier Metal Company Limited)

Material	Fatigue rating[1]		Seizure load[2]	
	MN/m^2	lbf/in^2	MN/m^2	lbf/in^2
Tin-base white metal	35	5000	14	2000 not seized
Lead-base white metal	35	5000	14	2000 not seized
70/30 copper–lead, unplated	95	13 500	11–14	1600–2000
70/30 copper–lead, overlay plated	70 119	10 000 overlay 17 000 copper–lead	—[3] —[3]	—[3] —[3]
Lead bronze (22% lead, 4% tin), unplated	—	—	5.5–11	800–1600
Lead bronze (22% lead, 4% tin), overlay plated	70 125	10 000 overlay 18 000 lead bronze	—[3]	—[3]
Lead bronze (10% lead, 10% tin), unplated	—	—	3–8.5	400–1200
6% tin–aluminium, unplated	105	15 000	5.5–14	800–2000
6% tin–aluminium, overlay plated	76 114	11 000 overlay 16 500 tin–aluminium	—[3] —[3]	—[3] —[3]
Aluminium–tin–silicon	118	13000	10–14	1500–2000
Aluminium–20% tin	90	13000	14	2000 usually

Notes: (1) Fatigue ratings determined on single-cylinder test rig. Not to be used for engine design purposes.
(2) Seizure load determined by stop–start tests on bushes. Maximum load on rig 14 MN/m^2 (2000 lbf/in^2).
(3) Overlay does not seize, but wears away. Seizure then occurs between interlayer and journal at load depending upon thickness of overlay, i.e. rate of wear. The overlay thickness on aluminium–tin is usually less than that on copper–lead and lead bronze, hence the slightly higher fatigue rating.

Characteristics of rubbing bearing materials

Material	Maximum P loading		PV value		Maximum temperature °C	Coefficient of friction	Coefficient of expansion ×10⁻⁶/°C	...ments	Application
	lbf/in²	MN/m²	lbf/in² × ft/min	MN/m² × m/s					
Carbon/graphite	200–300	1.4–2	≯3000 for continuous operation 5000 for short period life	0.11 for continuous operation 0.18 for short period life	350–500	0.10–0.25 dry	2.5–5.0	For continuous dry operation $P \not> 200$ lbf/in² $(1.4\ \mathrm{MN/m^2})$, $V \not> 250$ ft/min $(1.25\ \mathrm{m/s})$	Food and textile machinery where contamination by lubricant inadmissible; furnaces, conveyors, etc. where temperature too high for conventional lubricants; where bearings are immersed in liquids, e.g. water, acid or alkaline solutions, solvents, etc.
Carbon/graphite with metal	450–600	3–4	4000 for continuous operation 6000 for short period	0.145 for continuous operation 0.22 for short period	130–350	0.10–0.35 dry	4.2–5.0	Permissible peak load and temperature depend upon metal impregnant	Bearings working in dusty atmospheres, e.g. coal-mining, foundry plant, steel plant, etc.
Graphite impregnated metal	10 000	70	8000–10 000	0.28–0.35	350–600	0.10–0.15 dry 0.020–0.025 grease lubricated	12–13 with iron matrix 16–20 with bronze matrix	Operates satisfactorily dry within stated limits; benefits considerably if small quantity of lubricant present, i.e. higher PV values	
Graphite/thermosetting resin	300	2	~10 000	~0.35	250	0.13–0.5 dry	3.5–5.0	Particularly suitable for operation in sea-water or corrosive fluids	
Reinforced thermosetting plastics	5000	35	~10 000	~0.35	200	0.1–0.4 dry 0.006 claimed with water lubrication	25–80 depending on plane of reinforcement	Values depend upon type of reinforcement, e.g. cloth, asbestos, etc. Higher PV values when lubricated	Water-lubricated roll-neck bearings (esp. hot rolling mills), marine sterntube and rudder bearings; bearings subject to atomic radiation
Thermo-plastic material without filler	1500	10	~1000	~0.035	100	0.1–0.45 dry	~100	Higher PV values acceptable if higher wear rates tolerated. With initial lubrication only, PV values up to 20 000 can be imposed	Bushes and thrust washers in automotive, textile and food machinery—linkages where lubrication difficult

Characteristics of rubbing bearing materials (continued)

Material	Maximum P loading		PV value		Maximum temperature °C	Coefficient of friction	Coefficient of expansion ×10⁻⁶/°C	Comments	Applications
	lbf/in²	MN/m²	lbf/in² × ft/min	MN/m² × m/s					
Thermo-plastic with filler or metal-backed	1500–2000	10–14	1000–3000	0.035–0.11	100	0.15–0.40 dry	80–100	Higher loadings and PV values sustained by metal-backed components, especially if lubricated	As above, and for more heavily loaded applications
Thermo-plastic with filler bonded to metal back	20 000	140	10 000	0.35	105	0.20–0.35 dry	27	With initial lubrication only. PV values up to 40 000 acceptable with re-lubrication at 500–1000 h intervals	For conditions of intermittent operation or boundary lubrication, or where lubrication limited to assembly or servicing periods, e.g. ball-joints, suspension and steering linkages, king-pin bushes, gearbox bushes, etc.
Filled PTFE	1000	7	Up to 10 000	Up to 0.35	250	0.05–0.35 dry	60–80	Many different types of filler used, e.g. glass, mica, bronze, graphite. Permissible PV and unit load and wear rate depend upon filler material, temperature, mating surface material and finish	For dry operation where low friction and low wear rate required, e.g. bushes, thrust washers, slideways, etc., may also be used lubricated
PTFE with filler, bonded to steel backing	20 000	140	Up to 50 000 continuous rating	Up to 1.75	280	0.05–0.30 dry	20 (lining)	Sintered bronze, bonded to steel backing, and impregnated with PTFE/lead	Aircraft controls, linkages; automotive gearboxes, clutch, steering suspension, bushes, conveyors, bridge and building expansion bearings
Woven PTFE reinforced and bonded to metal backing	60 000	420	Up to 45 000 continuous rating	Up to 1.60	250	0.03–0.30 dry	—	The reinforcement may be interwoven glass fibre or rayon	Aircraft and engine controls, linkages, automotive suspensions, engine mountings, bridge and building expansion bearings

Notes: (1) Rates of wear for a given material are influenced by load, speed, temperature, material and finish of mating surface. The *PV* values quoted in the above table are based upon a wear rate of 0.001 in (0.025 × 10⁻³ m) per 100 h, where such data are available. For specific applications higher or lower wear rates may be acceptable—consult the bearing supplier.

(2) Where lubrication is provided, either by conventional lubricants or by process fluids, considerably higher *PV* values can usually be tolerated than for dry operation.

MATERIALS

Usually composites based on polymers, carbons, and metals.

The properties of typical dry rubbing bearing materials

Type	Examples	Max. static load		Max. service temp.	Coeff. exp.	Heat conductivity		Special features
		MN/m²	10³ lbf/in²	°C	10⁶/°C	W/m°C	Btu/ft h °F	
Thermoplastics	Nylon, acetal, UHMWPE	10	1.5	100	100	0.24	0.14	Inexpensive
Thermoplastics +fillers	Above+MoS_2, PTFE, glass, graphite, etc.	15–20	2–3	150	60–100	0.24	0.14	Solid lubricants reduce friction
PTFE+fillers	Glass, bronze, mica, carbon, metals	2–7	0.3–1	250	60–100	0.25–0.5	0.15–0.3	Very low friction
High temperature polymers (+fillers)	Polyimides polyamide-imide PEEK	30–80	4.5–12	250	20–50	0.3–0.7	0.2–0.4	Relatively expensive
Thermosets +fillers	Phenolics, epoxies +asbestos, textiles, PTFE	30–50	4.5–7.5	175	10–80	0.4	0.25	Reinforcing fibres improve strength
Carbon–graphite	Varying graphite content; may contain resin	1–3	0.15–0.45	500	1.5–4	10–50	6–30	Chemically inert
Carbon–metal	With Cu, Ag, Sb, Sn, Pb	3–5	0.45–0.75	350	4–5	15–30	9–18	Strength increased
Metal–solid lubricant	Bronze–graphite -MoS_2; Ag–PTFE	30–70	4.5–10	250–500	10–20	50–100	30–60	High temperature capability
Special non-machinable products	Porous bronze/ PTFE/Pb	350	50	275	20	42	24	Need to be considered at the design stage
	PTFE/glass weave+resin	700	100	250	12	0.24	0.14	
	Thermoset+ PTFE surface	50	7.5	150	10	0.3	0.2	
	Metal+filled PTFE liner	7	1	275	100	0.3	0.2	

Notes:
   All values are approximate; properties of many materials are anisotropic.
   Most materials are available in various forms: rod, sheet, tube, etc.
   For more detailed information, consult the supplier, or ESDU Data Item 87007.

EFFECT OF ENVIRONMENT

Type of material	Temp. above 200°C	Temp. below −50°C	Radiation	Vacuum	Water	Oils	Abrasives	Acids and alkalis
Thermoplastics +fillers	Few suitable	Usually good	Usually poor	Most materials suitable; avoid graphite as fillers	Often poor; watch finish of mating surface	Usually good	Poor to fair; rubbery materials best	Fair to good
PTFE+fillers	Fair	Very good	Very poor					Excellent
Thermosets+ fillers	Some suitable	Good	Some fair					Some good
Carbon– graphite	Very good; watch resins and metals	Very good	Very good; avoid resins	Useless	Fair to good	Good	Poor	Good, except strong acids

PERFORMANCE

Best criterion of performance is a curve of P against V for a specified wear rate. The use of $P \times V$ factors can be misleading.

Curves relate to journal bearings with a wear rate of 25 μm (1 thou.)/100 h—unidirectional load; 12.5 μm (0.5 thou.)/100 h—rotating load
Counterface finish 0.2–0.4 μm cla (8–16 μin).

A   Thermoplastics
B   PTFE
C   PTFE+fillers
D   Porous bronze+PTFE+Pb
E   PTFE–glass weave+thermoset
F   Reinforced thermoset+MoS$_2$
G   Thermoset/carbon-graphite+PTFE

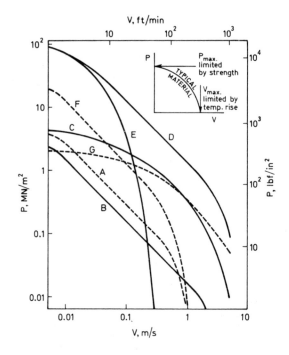

WEAR

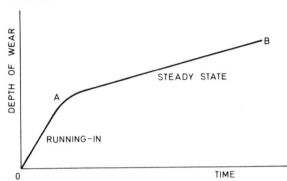

'Running-in' wear O–A is very dependent upon counterface roughness. Approximately, wear rate α (cla roughness)

'Steady-state' wear A–B depends on (*i*) mechanical properties of the material and its ability to (*ii*) smooth the counterface surface and/or (*iii*) transfer a thin film of debris.

In general, the steady-state wear rate, depth/unit time = KPV (*a.b.c.d.e*). K is a material constant incorporating (*i*), (*ii*), and (*iii*) above. Wear-rate correction factors *a,b,c,d,e*, depend on the operating conditions as shown below.

Approximate values of wear-rate correction factors

a, Geometrical factor
- continuous motion
  - rotating load — 0.5
  - unidirectional load — 1
- oscillatory motion — 2

b, Heat dissipation factor
- metal housing, thin shell, intermittent operation — 0.5
- metal housing, continuous operation — 1
- non-metallic housing, continuous operation — 2

c, Temperature factor

	PTFE–base	carbon–graphite, thermosets
20°C	1	1
100°C	2	3
200°C	5	6

d, Counterface factor
- stainless steels, chrome plate — 0.5
- steels — 1
- soft, non-ferrous metals (Cu alloys, Al alloys) — 2–5

e, Surface finish factor
- 0.1–0.2 μm cla (4–8 μin) — 1
- 0.2–0.4 μm cla (8–16 μin) — 2–3
- 0.4–0.8 μm cla (16–32 μin) — 4–10

Note: Factors do not apply to metal–solid lubricant composites.

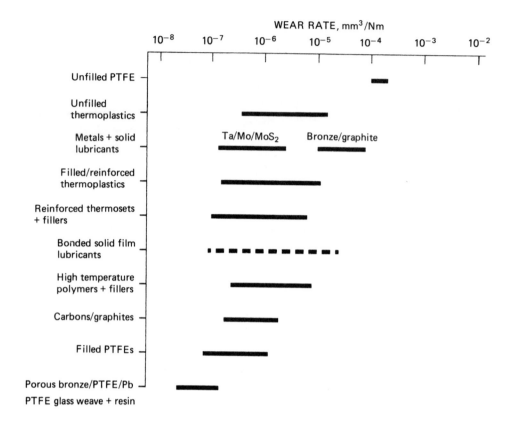

Order-of-magnitude wear rates of dry bearing material groups. At light loads and low speeds (frictional heating negligible) against smooth (0.15 μm Ra) mild steel

POINTS TO NOTE IN DESIGN

Choose length/diameter ratio between $\frac{1}{2}$ and $1\frac{1}{2}$.
Minimise wall thickness to aid heat dissipation.

Possibility of dimensional changes after machining
$\begin{cases} \text{moisture absorption} \\ \text{high expansion coefficients} \\ \text{stress relaxation} \end{cases}$

Machining tolerances may be poor: 25–50 μm (1–2 thou.) for plastics; better for carbons.

Suitable housing location methods are
$\begin{cases} \text{plastics—mechanical interlock or adhesives} \\ \text{metal-backed plastics—interference fit} \\ \text{carbon-graphite—press or shrink fit} \end{cases}$

Avoid soft shafts if abrasive fillers present, e.g. glass.
Minimise shaft roughness: 0.1–0.2 μm cla (4–8 μin) preferred.

Allow generous running clearances
$\begin{cases} \text{plastics, 5 μm/mm (5 thou./in). min. 0.1 mm (4 thou.)} \\ \text{carbon-graphite, 2 μm/mm (2 thou./in). min, 0.075 mm (3 thou.)} \end{cases}$

Contamination by fluids, or lubrication, usually lowers friction but:

increases wear of filled PTFE's and other plastics containing PTFE, graphite or MoS_2;
decreases wear of thermoplastics and thermosets without solid lubricant fillers.

DESIGN AND MATERIAL SELECTION

Having determined that a self-lubricating porous metal bearing may be suitable for the application, use Fig. 6.1 to assess whether the proposed design is likely to be critical for either load capacity or oil replenishment. With flanged bearings add together the duty of the cylindrical and thrust bearing surfaces.

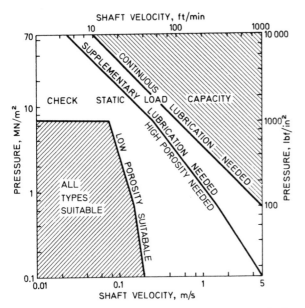

Fig. 6.1. A general guide to the severity of the duty. *At high pressures and particularly high velocities the running temperature increases, which requires provision for additional lubrication to give a satisfactory life. Attention to the heat conductivity of the assembly can reduce the problem of high running temperatures. High porosity bearings contain more oil but have lower strength and conductivity. The data are based on a length to diameter ratio of about 1, and optimisation of the other design variables*

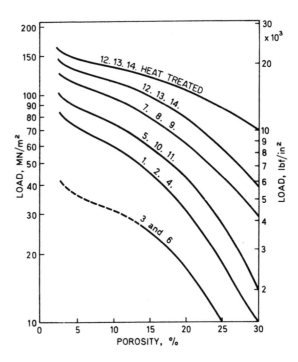

Fig. 6.2. A general guide to the maximum static load capacity (including impact loads) of a wide range of compositions and porosities. *The curves are based on a length to diameter of about 1, and assume a rigid housing. Note that all compositions are not available in all porosities and sizes*

Bearing strength

Figure 6.2 give the relationship between the maximum static load capacity and porosity for the fourteen different standard compositions listed in Table 6.1. Wherever possible select one of these preferred standards for which the design data in Fig. 6.3 and 6.4 apply. Having made the choice, check with the manufacturers that at the wall thickness and length-to-diameter ratio, the static load capacity is acceptable.

Wall thickness, L/d ratio, tolerances

The length, diameter and composition determine the minimum wall thickness which can be achieved, and avoid a very large porosity gradient in the axial direction. Porosity values are quoted as average porosity, and the porosity at the ends of the bearing is less than in the centre. As most properties are a function of the porosity, the effect of the porosity gradient on the performance has to be separately considered. The dimensional tolerances are also a function of the porosity gradient, wall thickness, length-to-diameter ratio, composition, etc.

Figure 6.3(a) gives the general case, and manufacturers publish, in tabular form, their limiting cases. A summary of these data is given in Fig. 6.4 for cylindrical and flanged bearings in the preferred standard composition and porosities indicated in Table 6.1. Clearly the problem is a continuous one, hence, when dealing with a critical design, aim for L/d about unity and avoid the corners of the stepped relationship in Fig. 6.4.

The corresponding limiting geometries and tolerances for thrust bearings and self-aligning bearings are given in Figs 6.3(b) and 6.3(c). In all cases avoid the areas outside the enclosed area.

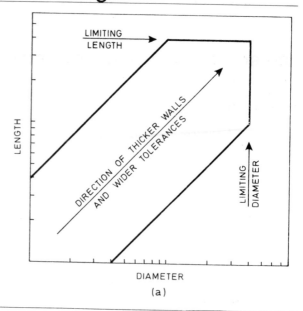

Fig. 6.3a. General effect of length and diameter on the minimum wall thickness and dimensional tolerance. *The stepped relationships present in Fig. 6.4 arise from a tabular interpretation of the continuous effect shown in Fig. 6.3a*

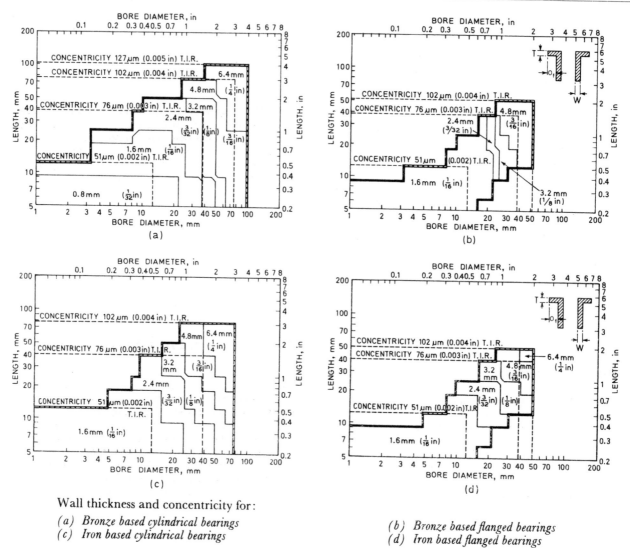

Wall thickness and concentricity for:

(a) *Bronze based cylindrical bearings*
(c) *Iron based cylindrical bearings*

(b) *Bronze based flanged bearings*
(d) *Iron based flanged bearings*

Fig. 6.4. Recommended minimum wall thicknesses and standard tolerances of diameters, length and concentricity for bearings to the preferred standards in Table 6.1.

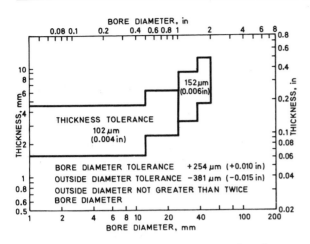

Fig. 6.3b. The thickness of thrust bearings or washers is a function of their diameter, as shown by the area bounded above

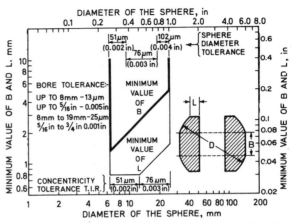

Fig. 6.3c. Self aligning bearings are standard between $\frac{1}{4}''$ and $1''$ sphere diameter. *Within this range the minimum length of the flat on the end faces (L) and the sphere diameter (B) is given above, together with concentricity, bore, and sphere diameter tolerances*

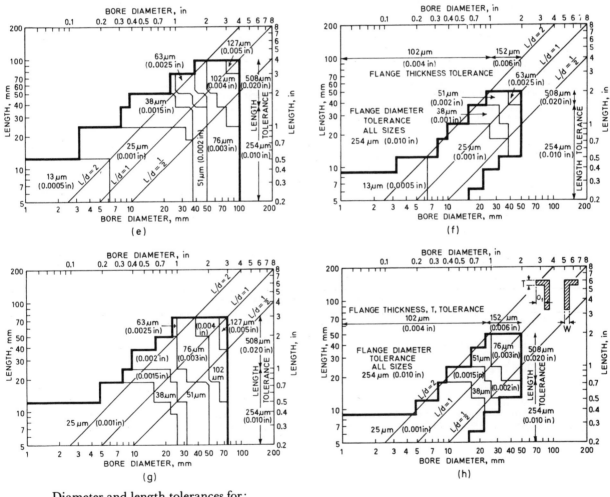

Diameter and length tolerances for:

(e) Bronze based cylindrical bearings
(g) Iron based cylindrical bearings

(f) Bronze based flanged bearings
(h) Iron based flanged bearings

Note: (1) For flanged bearings these data apply only where $T \simeq W$ and where $T \leqslant 0_f \leqslant 3\,T$.
(2) Smaller tolerance levels can usually be supplied to special order.
(3) Data for other compositions or porosities are available from the manufacturers.

Porous metal bearings

Composition and porosity

The graphited tin bronze (No.1 in Table 6.1) is the general purpose alloy and gives a good balance between strength, wear resistance, conformability and ease of manufacture. Softer versions have lead (No. 4) or reduced tin (No. 2). Graphite increases the safety factor if oil replenishment is forgotten, and the high graphite version (No. 3) gives some dry lubrication properties at the expense of strength.

Where rusting is not a problem, the cheaper and stronger iron-based alloys can be used. Soft iron (No. 5) has a low safety factor against oil starvation, especially with soft steel shafts. Graphite (Nos. 6 and 10) improves this, but reduces the strength unless the iron is carburised during sintering (No. 11). Copper (Nos. 7, 8 and 9) increases the strength and safety factor. If combined with carbon (Nos. 12, 13 and 14) it gives the greatest strength especially after heat treatment.

Table 6.1 Typical specifications for porous metal bearing materials

No. ref. Fig. 6.2	Composition	Notes on composition
1	**89/10/1 Cu/Sn/graphite**	General purpose bronze (normally supplied unless otherwise specified). Reasonably tolerant to unhardened shafts
2	91/8/1 Cu/Sn/graphite	Lower tin bronze. Reduced cost. Softer
3	85/10/5 Cu/Sn/graphite	High graphite bronze. Low loads. Increased tolerance towards oil starvation
4	86/10/3/1 Cu/Sn/Pb/graphite	Leaded bronze. Softer. Increased tolerance towards misalignment
5	>99% iron (soft)	Soft iron. Cheaper than bronze. Unsuitable for corrosive conditions. Hardened shafts preferred
6	$97\frac{1}{2}/2\frac{1}{2}$ Fe/graphite	Graphite improves marginal lubrication and increases tolerance towards unhardened shafts
7	98/2 Fe/Cu	Increasing copper content increases strength and cost. This series forms the most popular range of porous iron bearings. Hardened shafts preferred
8	**2% to 25% Cu in Fe**	
9	75/25 Fe/Cu	
10	89/10/2 Fe/Cu/graphite	High graphite improves marginal lubrication and increases tolerance towards unhardened shafts
11	99/0.4 Fe/C	Copper free, hardened steel material
12	97/2/0.7 Fe/Cu/C	Hardened high strength porous steels. Increasing copper content gives increasing strength and cost
13	2% to 10% Cu in 0.7 C/Fe	
14	89/10/0.7 Fe/Cu/C	

Note: These typical specifications are examples of materials listed in various relevant standards such as: ISO 5755/1, BS 5600/5/1, DIN 30 910/3, MPIF/35, ASTM B 438, ASTM 439. Most manufacturers offer a wide choice of compositions and porosities.

LUBRICATION

As a general recommendation, the oil in the pores should be replenished every 1000 hours of use or every year, whichever is the sooner. However, the data in Fig. 6.5 should be used to modify this general recommendation. Low porosity bearings should be replenished more frequently. Bearings running submerged or receiving oil-splash will not require replenishment. See the notes in Table 6.1 about compositions which are more tolerant to oil starvation. Figure 6.6 gives details of some typical assemblies with provision for supplementary lubrication.

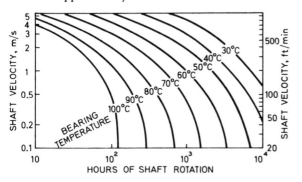

Fig. 6.5. The need to replenish the oil in the pores arises because of oil loss (which increases with shaft velocity) and oil deterioration (which increases with running temperature). *The above curves relate to the preferred standard bearing materials in Table 6.1*

Selection of lubricant

1 Figure 6.7 gives general guidance on the choice of oil viscosity according to load and temperature.
2 Lubricants must have high oxidation resistance.
3 Unless otherwise specified, most standard porous metal bearings are impregnated with a highly refined and oxidation-inhibited oil with an SAE 20/30 viscosity.
4 Do not select oils which are not miscible with common mineral oils unless replenishment by the user with the wrong oil can be safeguarded.
5 Do not use grease, except to fill a blind cavity of a sealed assembly (see Fig. 6.6).
6 Avoid suspensions of solid lubricants unless experience in special applications indicates otherwise.
7 For methods of re-impregnation—consult the manufacturers.

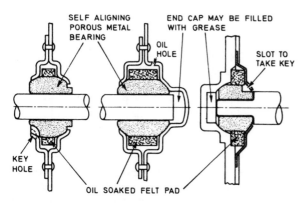

ASSEMBLIES OF SELF-ALIGNING POROUS METAL BEARINGS WITH PROVISION FOR ADDITIONAL LUBRICATION

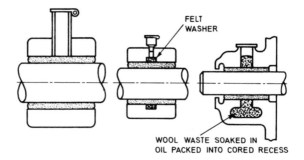

WOOL WASTE SOAKED IN OIL PACKED INTO CORED RECESS

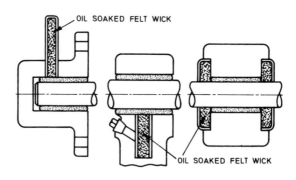

TYPICAL METHODS OF SUPPLEMENTING AND REPLENISHING THE OIL IN THE PORES OF A FORCE FITTING BEARING

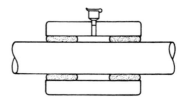

SIMPLE LUBRICATION ARRANGEMENT WHICH CAN BE EMPLOYED WITH A PAIR OF FORCE FITTED POROUS METAL BEARINGS

Fig. 6.6. Some typical assemblies showing alternative means of providing supplementary lubrication facilities

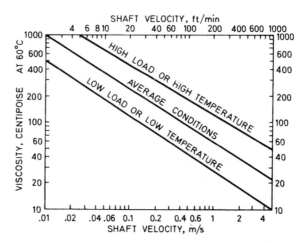

Fig. 6.7. General guide to the selection of oil viscosity expressed in centipoises at 60°C

INSTALLATION

General instructions

1 Ensure that the bearings are free of grit, and wash in oil if not held in dust-free storage. Re-impregnate if held in stock for more than one year or if stored in contact with an oil absorbent material.

2 With a self-aligning assembly (see example in Fig. 6.6):
  (a) ensure that the sphere is able to turn freely under the action of the misalignment force;
  (b) check that the static load capacity of the housing assembly is adequate;
  (c) note that the heat dissipation will be less than a force-fitted assembly and hence the temperature rise will be higher.

3 With a force-fitted assembly (see examples in Fig. 6.6):
  (a) select a mean diametral interference of $0.025 + 0.0075\sqrt{D}$ mm $(0.001 + 0.0015\sqrt{D}$ inches);
  (b) check that the stacking of tolerances of housing and bearing (see Fig. 6.4) keeps the interference between about half and twice the mean interference;
  (c) allow adequate chamfer on the housing (see Table 6.2 for details);
  (d) estimate the bore closure on fitting using the F factor from Fig. 6.8 and the extremes of interference from (b) above. Select a fitted bore size which is not smaller than 'the unfitted bore size minus the bore closure'. Check at the extremes of the tolerances of interference and bore diameter (see Fig. 6.4);
  (e) estimate the diameter of the fitting mandrel shown in Fig. 6.9, by adding to the desired bore size, a spring allowance which varies with the rigidity of the porous metal (Fig. 6.2) and the housing, as given in Table 6.3;
  (f) check that the differential thermal expansion between the housing and bearing over the expected temperature range does not cause a loss of interference in service (use the expansion coefficient of a non-porous metal of the same composition for all porosities);
  (g) for non-rigid housings, non-standard bearings or where the above guidance does not give a viable design, consult the manufacturers.

4 Never use hammer blows, as the impact force will generally exceed the limiting load capacity given in Fig. 6.2. A steady squeezing action is recommended.

5 Select a mean running clearance from Fig. 6.10, according to shaft diameter and speed. Check that the stacking of tolerances and the differential expansion give an acceptable clearance at the extremes of the design. Note that excessive clearance may give noisy running with an out-of-balance load, and that insufficient clearance gives high torque and temperature.

6 Specify a shaft-surface roughness of about 0.8 μm (32 micro-inches) cla, remembering that larger diameters can tolerate a greater roughness, and that a smaller roughness gives better performance and less running-in debris. In critical applications (Fig. 6.1), iron based bearings using steel shafts need a smoother shaft finish than bronze based bearings.

Table 6.2 Minimum housing chamfers at 45°

Housing diameter, D	Length of chamfer
Up to 13 mm ($\frac{1}{2}$ in)	0.8 mm ($\frac{1}{32}$ in)
13 mm to 25 mm ($\frac{1}{2}$ in to 1 in)	1.2 mm ($\frac{3}{64}$ in)
25 mm to 51 mm (1 in to 2 in)	1.6 mm ($\frac{1}{16}$ in)
51 mm to 102 mm (2 in to 4 in)	2.4 mm ($\frac{3}{32}$ in)
Over 102 mm (4 in)	3.2 mm ($\frac{1}{8}$ in)

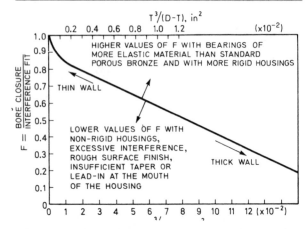

Fig. 6.8. Ratio of interference to bore closure, F, as a function of the wall thickness, W, and outside diameter, D, of the porous metal bearing

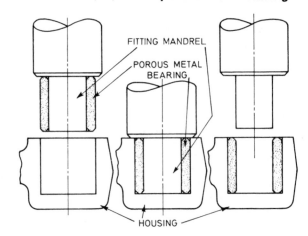

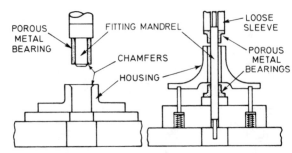

Fig. 6.9. Force fitting of porous metal bearings using a fitting mandrel to control the fitted bore diameter and to achieve alignment of a pair of bearings

Table 6.3 Spring allowance on force fitting mandrel

Static load capacity (Fig. 6.2)		Spring allowance
Up to 20 MN/m²	(3 000 p.s.i.)	0.01%
20 to 40 MN/m²	(6 000 p.s.i.)	0.02%
40 to 80 MN/m²	(12 000 p.s.i.)	0.04%
Over 80 MN/m²	(12 000 p.s.i.)	0.06%

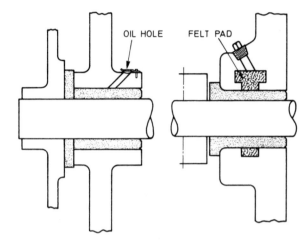

THRUST LOADS ARE CARRIED BY EITHER A SEPARATE THRUST WASHER OR THE USE OF A FLANGED CYLINDRICAL BEARING

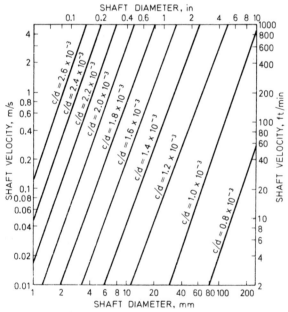

Fig. 6.10. Guide to the choice of mean diametral clearance expressed as the clearance ratio c/d

Bore correction

If after fitting, it is necessary to increase the bore diameter, this should not be done by a cutting tool, which will smear the surface pores and reduce the free flow of oil from the porosity to the working surface. Suitable burnishing tools for increasing the bore diameter, which do not close the pores, are given in Fig. 6.11.

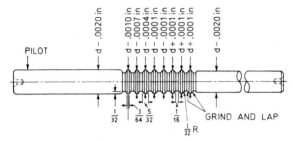

d = FINISHED DIAMETER OF BEARING

BUTTON TYPE DRIFT

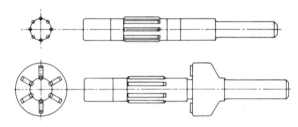

ROLLER TYPE BURNISHING TOOL

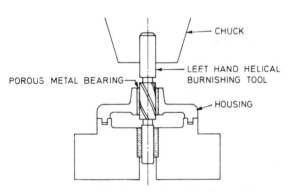

Fig. 6.11. Tools for increasing the bore diameter and aligning a fitted assembly

GENERAL NOTE

The previous sections on design, materials and lubrication give general guidance applicable to normal operating conditions with standard materials, and therefore cover more than half of the porous metal bearings in service. There are, however, many exceptions to these general rules, and for this reason the manufacturers should be consulted before finalising an important design.

Journal bearings lubricated with grease, or supplied with oil by a wick or drip feed, do not receive sufficient lubricant to produce a full load carrying film. They therefore operate with a starved film as shown in the diagram:

As a result of this film starvation, these bearings operate at low film thicknesses.

To make an estimate of their performance it is, therefore, necessary to take particular account of the bearing materials and the shaft and bearing surface finishes as well as the feed rate from the lubricant feed system.

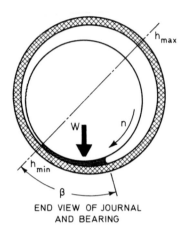

END VIEW OF JOURNAL
AND BEARING

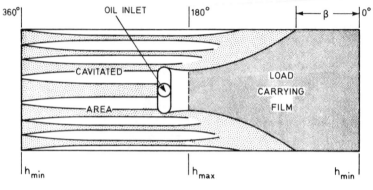

SWEPT AREA OF BEARING

STARVED FILM

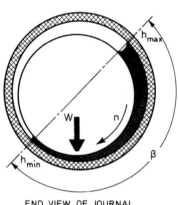

END VIEW OF JOURNAL
AND BEARING

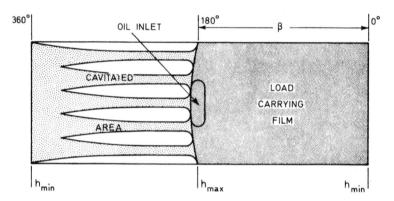

SWEPT AREA OF BEARING

FULL FILM

(AS OBTAINED WITH A PRESSURE OIL FEED)

AN APPROXIMATE METHOD FOR THE DESIGN OF STARVED FILM BEARINGS

Step 1

Check the suitability of a starved film bearing for the application using Fig. 7.1.

Note: in the shaded areas attention should be paid to surface finish, careful running-in, good alignment and the correct choice of materials for bearing and journal.

Bearing width to diameter ratio, b/d, should be between 0.7 and 1.3.

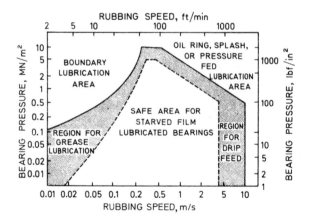

Fig. 7.1. A guide to the suitability of a 'starved' bearing

Step 2

Select a suitable clearance C_d, knowing the shaft diameter (Fig. 7.2) and the manufacturing accuracy.

Note: the lowest line in Fig. 7.2 gives clearance suitable only for bearings with excellent alignment and manufacturing precision.

For less accurate bearings, the diametral clearance should be increased to a value in the area above the lowest line by an amount, $= Mb +$ the sum of out-of-roundness and taper on the bearing and journal.

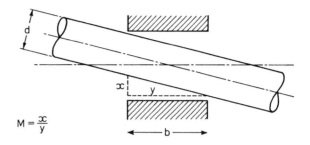

$M = \dfrac{x}{y}$

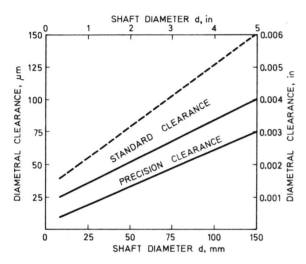

Fig. 7.2. Guidance on choice of clearance

Table 7.2 Surface finish, predominant peak height, R_p

Surface type	Micro-inch cla	μm RMS	Class	R_p μm	R_p μin
Turned or rough ground	100	2.8	6	12	480
Ground or fine bored	20	0.6	8	3	120
Fine ground	7	0.19	10	0.8	32
Lapped or polished	1.5	0.04	12	0.2	8

Step 4

Assume a lubricant running-temperature of about 50 to 60°C above ambient and choose a type and grade of lubricant with references to Tables 7.3 and 7.4. Note the viscosity corresponding to this temperature from Fig. 7.3.

Step 3

Choose the minimum permissible oil film thickness h_{min} corresponding to the materials, the surface roughnesses and amount of misalignment of the bearing and journal.

Minimum oil film thickness

$$h_{min} = k_m \left(R_p \text{ journal} + R_p \text{ bearing}\right) + \frac{Mb}{2}$$

Table 7.1 Material factor, k_m

Bearing lining material	k_m
Phosphor bronze	1
Leaded bronze	0.8
Tin aluminium	0.8
White metal (Babbitt)	0.5
Thermoplastic (bearing grade)	0.6
Thermosetting plastic	0.7

Note: journal material hardness should be five times bearing hardness.

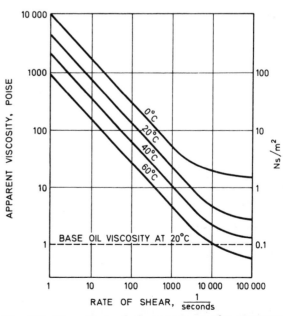

Fig. 7.3. The effect of shear rate on the apparent viscosity of a typical No. 2 NLGI consistency grease

Table 7.3 Guidance on the choice of lubricant grade

Lubricant running temperature	Grease			Oil	
	Type	Grade (NLGI No.)		Types	Viscosity grade ISO 3448
Up to 60°C	Calcium based 'cup grease'			Mineral oil with fatty additives	
< 0.5 m/s		1 or 2			68
> 0.5 m/s		0			32
60°C to 130°C	Lithium hydroxystearate based grease with high V.I. mineral oil and anti-oxidant additives			Good quality high V.I. crankcase or hydraulic oil with antioxidant additives (fatty oils for drip-fed bearings)	
< 0.5 m/s		3			150
> 0.5 m/s		3			68
Above 130°C	Clay based grease with silicone oil	3		Best quality fully inhibited mineral oil, synthetic oil designed for high temperatures, halogenated silicone oil	150

Notes: for short term use and total loss systems a lower category of lubricant may be adequate.
A lubricant should be chosen which contains fatty additives, i.e. with good 'oiliness' or 'lubricity'.
The use of solid lubricant additives such as molybdenum disulphide and graphite can help (but not where lubrication by wick is used).

Table 7.4 Factors to consider in the choice of grease as a lubricant

Feature	Advantage	Disadvantage	Practical effect
h_{min} Minimum film thickness	Fluid film lubrication maintained at lower W' values		Grease lubrication is better for high load, low-speed applications
C_d/d Clearance diameter ratio	Larger clearances are permissible	Overheating and feeding difficulties arise with small clearances	Ratios 2 to 3 times larger than those for oil lubricated bearings are common
Lubricant supply	Much smaller flow needed to maintain a lubricant film. Rheodynamic flow characteristics lead to small end-loss and good recirculation of lubricant	Little cooling effect of lubricant, even at high flow rates	Flow requirement 10 to 100 times less than with oil. Long period without lubricant flow possible with suitable design
μ Friction coefficient (a) at start-up (b) running	(a) Lubricant film persists under load with no rotation	(b) Higher effective viscosity leads to higher torque	(a) Lower start-up torque (b) Higher running temperatures
W' Bearing load capacity number		Calculated on the basis of an 'effective viscosity' value dependent on the shear rate and amount of working. Gives an approx. guide to performance only	Prediction of design performance parameters poor

Step 5

With reference to the formulae on Fig. 7.4 calculate W' from the dimensions and operating conditions of the bearing, using the viscosity just obtained. Obtain the appropriate misalignment factor M_w from Table 7.5. Calculate W' (misaligned) by multiplying by M_w. Use this value in further calculations involving W'.

Notes: M_w is the available percentage of the load capacity W' of a correctly aligned bearing.
Misalignment may occur on assembly or may result from shaft deflection under load.

Table 7.5 Values of misalignment factor M_w at two ratios of minimum oil film thickness/diametral clearance

$M \times b/c_d$	$h_{min}/C_d = 0.1$	$h_{min}/C_d = 0.01$
0	100	100
0.05	65	33
0.25	25	7
0.50	12	3
0.75	8	1

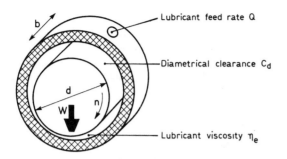

$Q' =$	$\dfrac{176.6\,Q}{bd\,n\,C_d}$	$\dfrac{2Q}{\pi\,bdn\,C_d}$
$W' =$	$\dfrac{4.137 \times 10^8\,W}{\eta_e\,nbd}\left(\dfrac{Cd}{d}\right)^2$	$\dfrac{W}{\eta_e\,nbd}\left(\dfrac{Cd}{d}\right)^2$
UNITS	gall/min inch lbf rev/min cP	m³/s m N rev/s Ns/m²

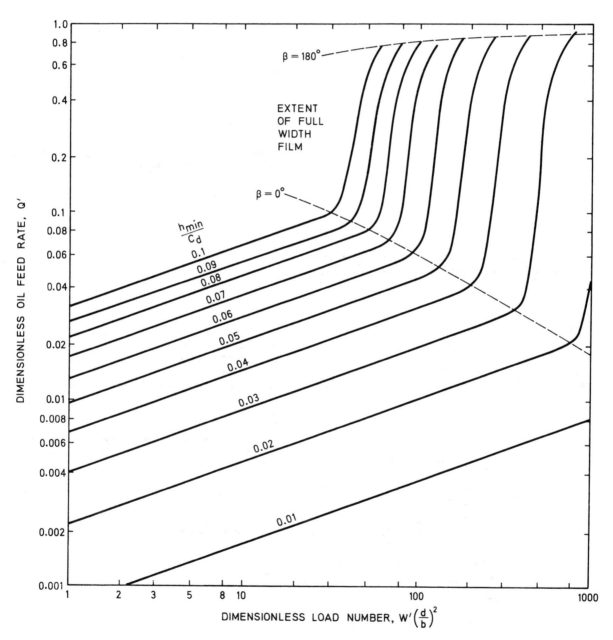

Fig. 7.4. Minimum oil flow requirements to maintain fluid film conditions, with continuous rotation, and load steady in magnitude and direction (*courtesy:* Glacier Metal Co Ltd)

Step 6

From Fig. 7.5 read the value of F' corresponding to this W'. Calculate the coefficient of fiction $\mu = F'C_d/d$.

Calculate the power loss H in watts

$H =$	$1.9 \times 10^{-3} \, \pi \mu \, W \, d \, n$	$\pi \mu \, W \, d \, n$
Units	in lbf rev/min	m N rev/s

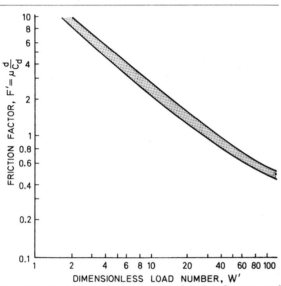

Fig. 7.5. Relationship between friction-factor and dimensionless load number (W' is defined on Fig. 7.4)

Step 7

It is assumed that all the power loss heat is dissipated from the housing surface. From Fig. 7.6 find the value of housing surface temperature above ambient which corresponds to the oil film temperature assumed in step 4. Read off the corresponding heat dissipation and hence derive the

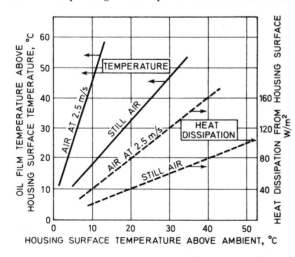

Fig. 7.6. A guide to the heat balance of the bearing housing

housing surface area using the power loss found in step 6. If this area is too large, a higher oil film temperature must be assumed and steps 4–7 repeated. It may be necessary to choose a different grade of lubricant to limit the oil film temperature.

Step 8

Using Fig. 7.4 read off Q' and calculate Q the minimum oil flow through the film corresponding to the dimensionless load number, $W' \times \left(\dfrac{d}{b}\right)^2$ and the value for h_{min}/C_d.

A large proportion of this flow is recirculated around the bearing and in each meniscus at the ends of the bearing.

An estimate of the required additional oil feed rate from the feed arrangement is given by $Q/10$ and this value may be used in step 9.

For grease lubrication calculate the grease supply rate per hour required Q_g from

$$Q_g = k_g \times C_d \times \pi \times d \times b$$

Table 7.6 Values of k_g for grease lubrication at various rotational speeds

Journal speed rev/min	k_g
Up to 100	0.1
250	0.2
500	0.4
1000	1.0

Under severe operating conditions such as caused by running at elevated temperatures, where there is vibration, where loads fluctuate or where the grease has to act as a seal against the ingress of dirt from the environment, supply rates of up to ten times the derived Q_g value are used.

Step 9

Select a type of lubricant supply to give the required rated lubricant feed using Tables 7.7 and 7.8 and Figs. 7.7, 7.8 and 7.9.

Where the rate of lubricant supply to the bearing is known, Fig. 7.4 will give the load number corresponding to a particular h_{min}/C_d ratio. The suggested design procedure stages should then be worked through, as appropriate.

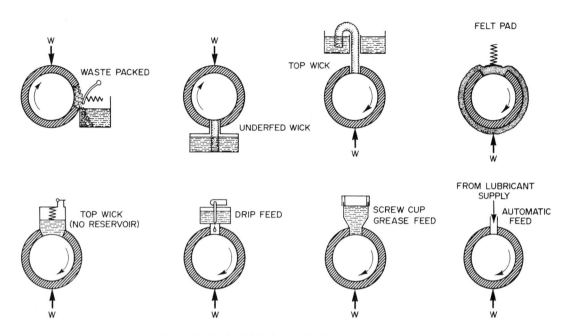

Fig. 7.7 Typical lubricant feed arrangements

Table 7.7 Guidance on the choice of lubricant feed system

Type	Lubricant supply method	Cost	Toleration of dirty environment	Maintenance needs rating	Lubricant flow characteristics
Wool waste lubricated	Capillary	Expensive housing design	Fair. Waste acts as an oil filter	Good. Infrequent refilling of reservoir	Very limited rate controlled by height of oil in reservoir. Recirculation possible. Varies automatically with shaft rubbing speed. Stops when rotation ceases
Wick lubricated (with reservoir)	Capillary and siphonic	Moderate	Fair. Wick acts as an oil filter	Good. Infrequent refilling of oil reservoir	Limited rate and control (ref. Fig. 7.8). Recirculation possible with underfed wick type. Varies slightly with shaft rubbing speed. Underfed type stops when rotation ceases (not siphonic)
Wick or pad lubricated (no reservoir)	Capillary	Cheap	Fair. Wick act as an oil filter	Fair. Reimpregnation needed occasionally	Very limited rate, decreasing with use. Varies slightly with shaft rubbing speed. Stops when rotation ceases. Recirculation possible
Grease lubricated	Hand-operated grease gun or screw cup	Very cheap	Good. Grease acts as a seal	Poor. Regular regreasing needed	Negligible flow, slumping only. Rheodynamic, i.e. no flow at low shear stress hence little end flow loss from bearing
Drip-feed lubricated	Gravity, through a controlled orifice	Cheap for simple installations	Poor	Poor. Regular refilling of reservoir needed	Variable supply rate. Constant flow at any setting. Total loss, i.e. no recirculation. Flow independent of rotation
Automatic feed (oil or grease)	Pump-applied pressure	Expensive ancillary equipment needed	Fair	Good. Supply system needs occasional attention	Wide range of flow rate. Can vary Automatically. Total loss. Can stop or start independently of rotation

Table 7.8 The comparative performance of various wick and packing materials

Type	Felt, high density (sg 3.4)	Felt, low density (sg 1.8 to 2.8)	Gilled thread	Wool waste	Cotton lamp wick
Height of oil lift (dependent on wetting and size of capillary channels)	Very good	Fair	Good	Poor	Fair
Rate of flow	Very good	Fair	Good	Poor	Fair
Oil capacity	Low	High	Low	Moderate (3 times weight of waste)	Fair
Suitability for use as packing	Poor (tendency to glaze)	Poor (tendency to glaze)	Poor	Good (superior elasticity)	Poor

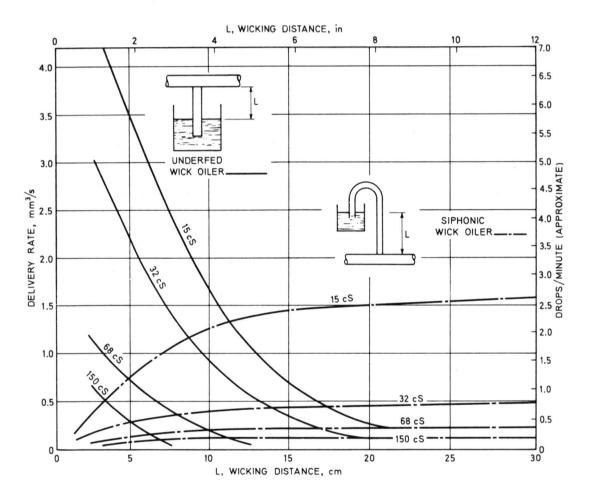

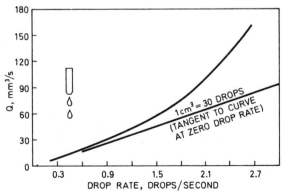

Fig. 7.8. Oil delivery rates for SAE F1 felt wicks, density 3.4g/cc, cross-sectional area 0.65cm² (0.1 in²), temperature 21°C, viscosity at 40°C (ISO 3448) (Data from the American Felt Co.)

Fig. 7.9. Effect of drop rate on oil drop size, temperature 27°C. Oil viscosity and lubricator tip shape have little effect on drop size over the normal working ranges

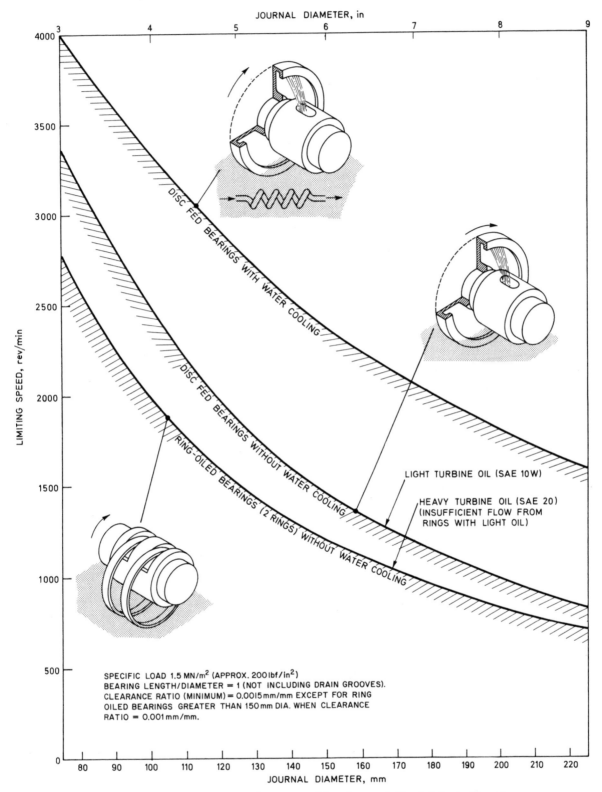

Fig. 8.1. General guide to limiting speed for ring and disc lubricated bearings

Disc fed—water cooled:

The above curves give some idea of what can be achieved, assuming there is sufficient oil to meet bearing requirement. It is advisable to work well below these limits. Typical maximum operating speeds used in practice are 75% of the above figures.

Ring and disc fed—without water cooling:

For more detailed information see Fig. 8.2. The limiting speed will be reduced for assemblies incorporating thrust location — see Fig. 8.5.

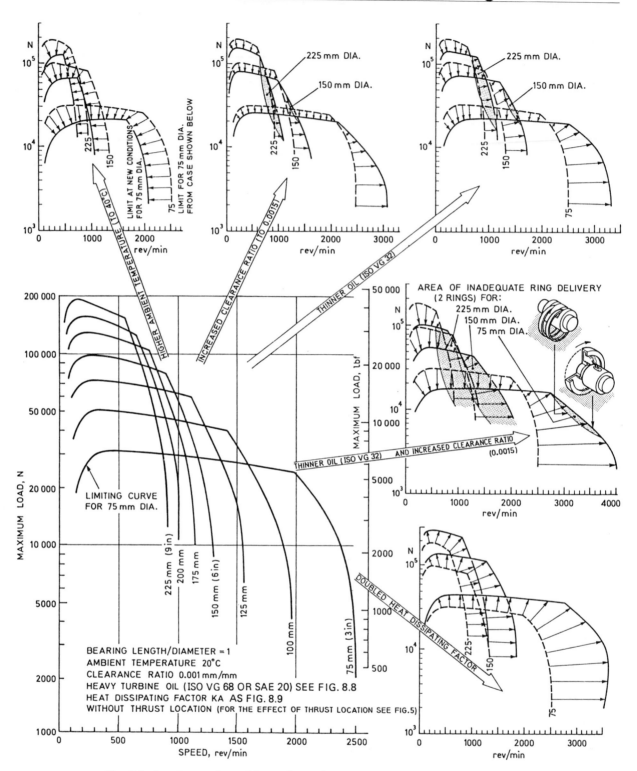

Fig. 8.2. Load capacity guidance for self-contained journal bearing assemblies

Disc fed:
For any diameter work below appropriate limiting curve*
(oil film thickness and temperature limits).

Ring oiled (2 rings):
For any diameter work below appropriate limiting curve*
and avoid shaded areas (inadequate supply of lubricant
from rings).

In these areas disc fed bearings should be used instead.

* These limits assume that the bearing is well aligned and adequately sealed against the ingress of dirt. Unless good alignment is achieved the load capacity will be severely reduced. In practice, the load is often restricted to 1.5 to 2 MN/m² (approx. 200 to 300 lbf/in²) to allow for unintentional misalignment, starting and stopping under load and other adverse conditions.

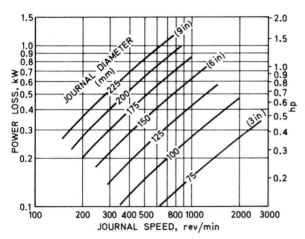

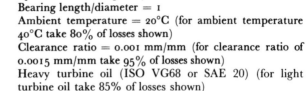

Fig. 8.3. *Guide for power loss in self-contained bearings*

Specific Load = 1.5 MN/m²
Bearing length/diameter = 1
Ambient temperature = 20°C (for ambient temperature 40°C take 80% of losses shown)
Clearance ratio = 0.001 mm/mm (for clearance ratio of 0.0015 mm/mm take 95% of losses shown)
Heavy turbine oil (ISO VG68 or SAE 20) (for light turbine oil take 85% of losses shown)
Heat dissipating factor as Fig. 8.9 (for effect of heat dissipating factor see Fig. 8.4)
The power loss will be higher for assemblies incorporating thrust location — see Fig. 8.6

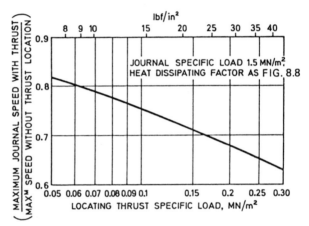

Fig. 8.5. *Reduced limiting speed where assembly includes thrust location*

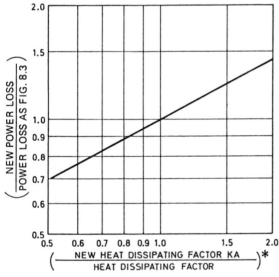

Fig. 8.4. *Showing how power loss in self-contained bearings (without thrust) is affected by heat dissipating factor KA*

The heat dissipating casing area A and/or the heat transfer coefficient K may both differ from the values used to derive the load capacity and power loss design charts.

Figure 8.4 shows how change in KA affects power loss.
*The ratio

$$\frac{\text{New heat dissipating factor } KA}{\text{Heat dissipating factor}} \text{ in Fig. 8.4}$$

is given by

$$\frac{K \text{ for actual air velocity (Fig. 8.9(}b\text{))}}{18 \text{ (for still air)}} \times \frac{\text{actual casing area}}{\text{casing area (Fig. 8.9(}a\text{))}}$$

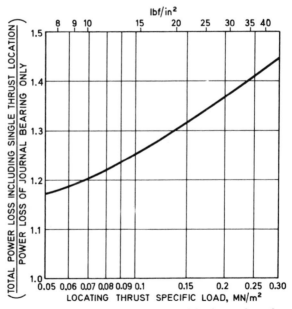

Fig. 8.6. *Increased power loss with thrust location (single thrust plain washer — for typical dimensions see Fig. 8.7)*

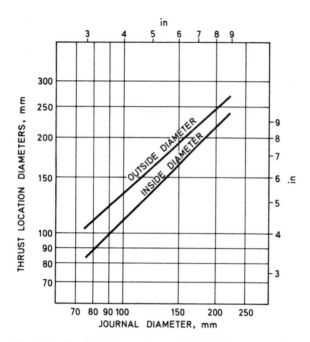

Fig. 8.7. Typical dimensions of plain thrust annulus as used in Figs. 8.5 and 8.6

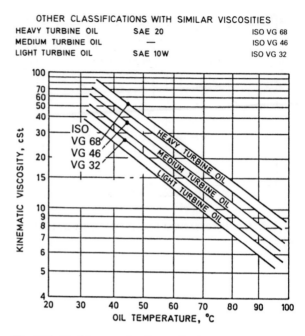

Fig 8.8. Turbine and other oil viscosity classifications

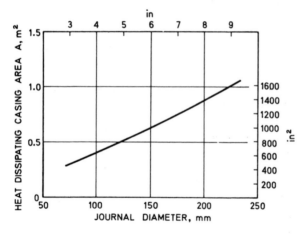

Fig. 8.9(a). Typical heat dissipating area of casing as used in the design guidance charts

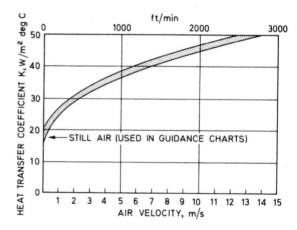

Fig. 8.9(b). Guidance on heat transfer coefficient K, depending on air velocity

The heat dissipating factor KA used in the design guidance charts was based on the area diameter relationship in Fig. 8.9(a) and a heat transfer coefficient for still air of 18 W/m^2 degC as shown in Fig. 8.9(b). The effect of different dissipating areas or air velocity over the casing may be judged:

    for load capacity Fig. 8.2 (doubled heat dissipating factor KA)
    for power loss Fig. 8.4.

HYDRODYNAMIC BEARINGS

Principle of operation

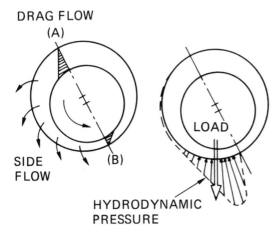

(1) On 'start up' the journal centre moves forming a converging oil film in the loaded region
(2) Oil is dragged into the converging film by the motion of the journal (see velocity triangle at 'A'). Similarly, a smaller amount passes through the minimum film (position 'B'*). As the oil is incompressible a hydrodynamic pressure is created causing the side flow
(3) The journal centre will find an equilibrium position such that this pressure supports the load

*The drag flow at these positions is modified to some extent by the hydrodynamic pressure.

Fig. 9.1 Working of hydrodynamic bearings – explained simply

The load capacity, for a given minimum film thickness increases with the drag flow and therefore increases with journal speed, bearing diameter and bearing length. It also increases with any resistance to side flow so will increase with operating viscosity. The bearing clearance may influence the load capacity either way. If the minimum film thickness is small and the bearing long then increasing the clearance could result in a decrease in load capacity, whereas an increase in clearance for a short bearing with a thick film could result in an increase in load capacity.

GUIDE TO PRELIMINARY DESIGN AND PERFORMANCE

The following guidance is intended to give a quick estimate of the bearing proportions and performance and of the required lubricant.

GUIDE TO GROOVING AND OIL FEED ARRANGEMENTS

An axial groove across the major portion of the bearing width in the unloaded sector of the bearing is a good supply method. A 2-axial groove arrangement, Fig. 9.2, with the grooves perpendicular to the loading direction is an arrangement commonly used in practice. The main design charts in this section relate to such a feed arrangement. A circumferential grooved bearing is used when the load direction varies considerably or rotates, but has a lower load capacity. However, with a 2-axial grooved bearing under small oil film thickness conditions, the load angle may be up to ±30° from the centre without significantly deviating the bearing. The lubricant is pumped into

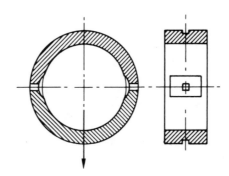

Fig. 9.2. Example of 2-axial groove bearing with load mid-way between grooves

the feed grooves at pressures from 0.07 to 0.35 MN/m².
0.1 MN/m² is used in the following design charts together with a feed temperature of 50°C.

BEARING DESIGN LIMITS

Figure 9.3 shows the concept of a *safe operating region* and Fig. 9.6 gives practical general guidance (also shows how the recommended operating region changes with different variables).

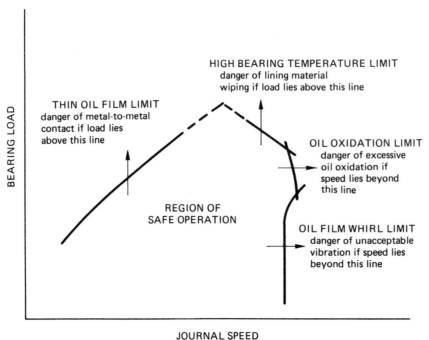

Fig. 9.3. Limits of safe operation for hydrodynamic journal bearings

Thin film limit – danger of metal to metal contact of the surfaces resulting in wear.

Background Safe limit taken as three times the peak-to-valley (R_{max}) value of surface finish on the journal. The factor of three, allowing for small unintentional misalignment and contamination of the oil is used in the general guide, Fig. 9.6. A factor of two may be satisfactory for very high standards of build and cleanliness. R_{max} depends on the trend in R_a values for different journal diameters as shown in Fig. 9.4 together with the associated machining process.

High bearing temperature limit – danger of bearing wiping at high speed conditions resulting in 'creep' or plastic flow of the material when subjected to hydrodynamic pressure. Narrow bearings operating at high speed are particularly prone to this limit.

Background The safe limit is well below the melting point of the bearing lining material. In the general guide, Fig. 9.6, whitemetal bearings are considered, with the bearing maximum temperature limited to 120°C. For higher temperatures other materials can be used: aluminium-tin (40% tin) up to 150°C and copper-lead up to 200°C. The former has the ability to withstand seizure conditions and dirt, and the latter is less tolerant so a thin soft overlay plate is recommended, togetjer with a hardened shaft and good filtration.

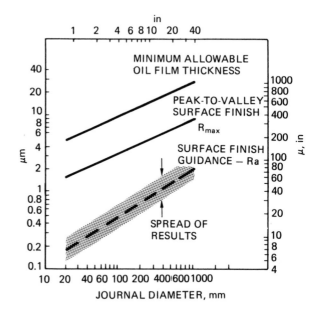

Fig. 9.4. Guidance on allowable oil film thickness dependent on surface finish

38

High temperature – oil oxidation limit – danger of excessive oil oxidation.
Background Industrial mineral oils can rapidly oxidize in an atmosphere containing oxygen (air). There is no precise limit; degradation is a function of temperature and operating period. Bulk drain temperature limit in the general guide. Fig. 9.6, is restricted to 75–80°C (assuming that the bulk temperatures of oil in tanks and reservoirs is of the same order).

Oil film whirl limit – danger of oil film instability.
Background Possible problem with lightly loaded bearings/rotors at high speeds.

RECOMMENDED MINIMUM DIAMETRAL CLEARANCE

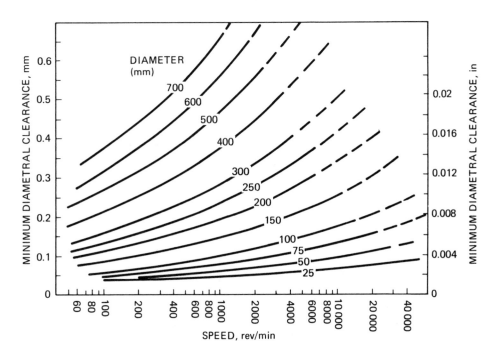

Fig. 9.5. Recommended minimum clearance for steadily loaded bearings
(dashed line region – possibility of non-laminar operation)

GUIDE FOR ESTIMATING MAXIMUM CLEARANCE

	Trends in bearing clearance tolerances – for bearing performance studies		
Bearing type	A bearing where the housing bore tolerance has little effect on bearing clearance (thickwalled or bored on assembly). Small tolerance range		A bearing which conforms to the housing bore. It has a larger tolerance as the wall thickness and housing bore must also be considered
Typical tolerance (mm) on diametral clearance	$\left(\dfrac{(\text{Bearing diameter, mm})^{1/3}}{80}\right)$	to	$\left(\dfrac{(\text{Bearing diameter, mm})^{1/3}}{60}\right)$

Maximum diametral clearance = Minimum recommended clearance (Fig. 9.5) + Tolerance (see trends above)

PRACTICAL GUIDE TO REGION OF SAFE OPERATION (INDICATING ACCEPTABLE GEOMETRY AND OIL GRADE)

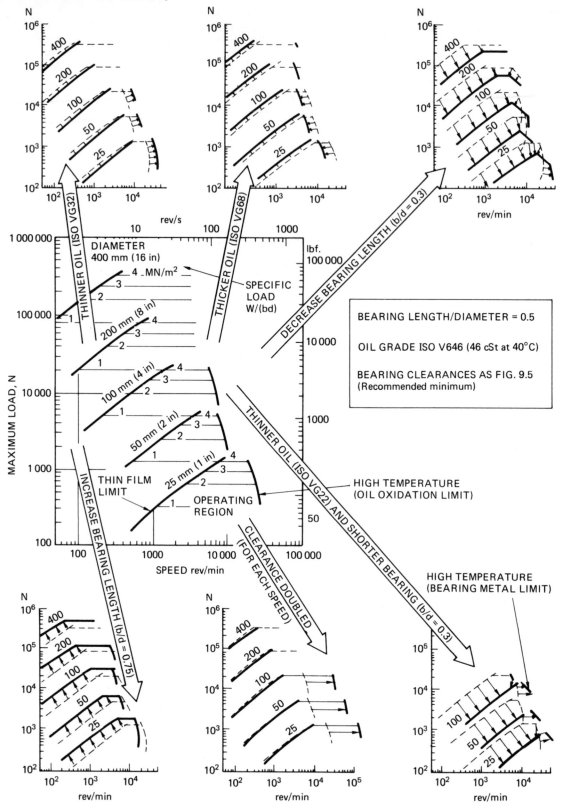

Fig. 9.6. *Guide to region of safe operation (showing the effect of design changes)* Work within the limiting curves

2 axial groove bearing – Groove length 0.8 of bearing length and groove width 0.25 of bearing diameter
Oil feed conditions at bearing – Oil feed pressure 0.1 MN/m² and oil feed temperature 50°C

BEARING LOAD CAPACITY

Operating load

The bearing load capacity is often quoted in terms of specific load (load divided by projected area of the bearing, W/bd) and it is common practice to keep the specific load below 4 MN/m². This is consistent with the practical guide shown in Fig. 9.6 which also shows that loads may have to be much lower than this in order to work within an appropriate speed range.

Guide to start-up load limit

For whitemetal bearings the start-up load should be limited to the following values:

	Specific load limit* at start-up	MN/m²
Frequent stops/starts	Several a day	1.4
Infrequent stops/starts	One a day or less	2.5

* Other limits at operating speeds must still be allowed for as shown in Fig. 9.6

BEARING PERFORMANCE

Figures 9.7 to 9.9 give the predicted minimum film thickness, power loss and oil flow requirements for a 2-axial grooved bearing with the groove geometry and feed conditions shown in Fig. 9.7. Any diametral clearance ratio C_d/d can be considered; however, the maximum should be used when estimating flow requirements. In some cases it may be necessary to judge the influence of different load line positions (at thick film conditions) or misalignment; both are considered in Figs. 9.10 to 9.12. A design guide 'check list' is given below.

DESIGN GUIDE CHECK LIST

(i) Using minimum clearance

	Information
See recommended minimum clearance	Fig. 9.5
Check that the bearing is within a safe region of operation adjust (or choose) geometry and/or oil as found necessary	Fig. 9.6
Predict oil film thickness ratio (minimum film thickness/diametral clearance)	Fig. 9.7
Predict power loss	Fig. 9.8
Allow for non-symmetric load angle (relative to grooves), if necessary	Figs 9.10 and 9.11
Allow for the influence of misalignment on film thickness, if necessary	Fig. 9.12
Check that modified minimum film thickness is acceptable	Fig. 9.4

(ii) Using maximum clearance

	Information
Calculate maximum clearance	Fig. 9.5 and tolerance equation
Predict film thickness and check that it is acceptable	Figs 9.7 and 9.4
Predict oil flow requirements	Fig. 9.9
Allow for non-symmetric load line and/or misalignment if necessary	Figs 9.10, 9.11 and 9.12
Check that modified minimum film thickness is acceptable	Fig. 9.4

GUIDE TO OPERATING MINIMUM FILM THICKNESS

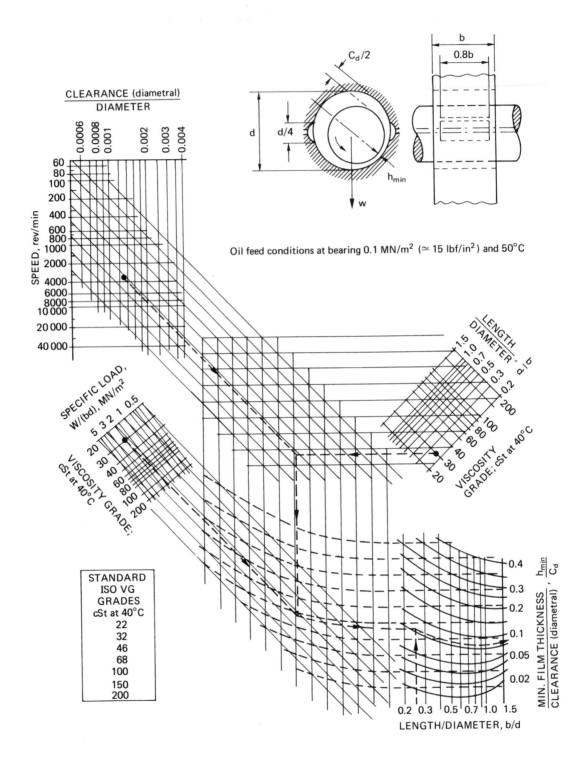

Oil feed conditions at bearing 0.1 MN/m² (≃ 15 lbf/in²) and 50°C

Fig. 9.7. Prediction of minimum oil film thickness for a centrally loaded bearing (mid-way between feed grooves) and for an aligned journal (laminar conditions)

GUIDE TO POWER LOSS

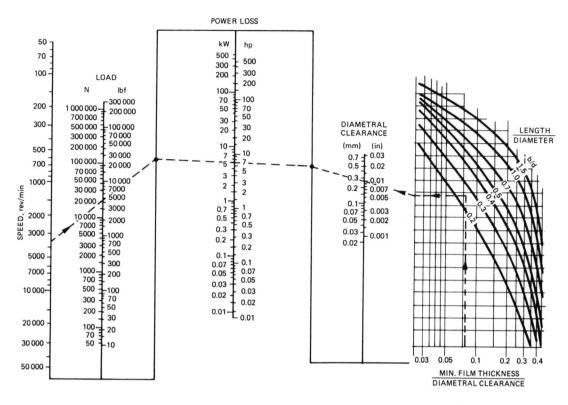

Fig. 9.8. Prediction of bearing power loss

GUIDE TO OIL FLOW REQUIREMENT

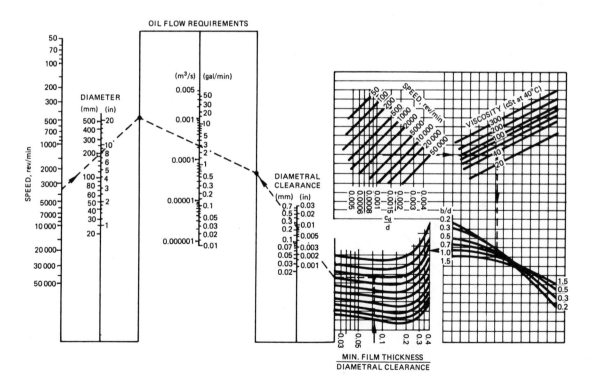

Fig. 9.9. Prediction of bearing oil flow requirement

EFFECT OF LOAD ANGLE ON FILM THICKNESS

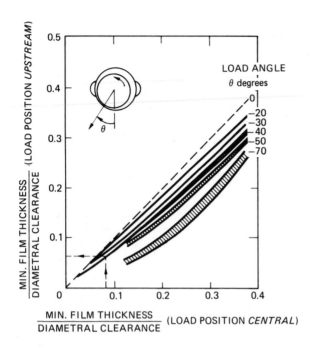

Fig. 9.10. Influence of load position on minimum film thickness (with load position upstream of centre)

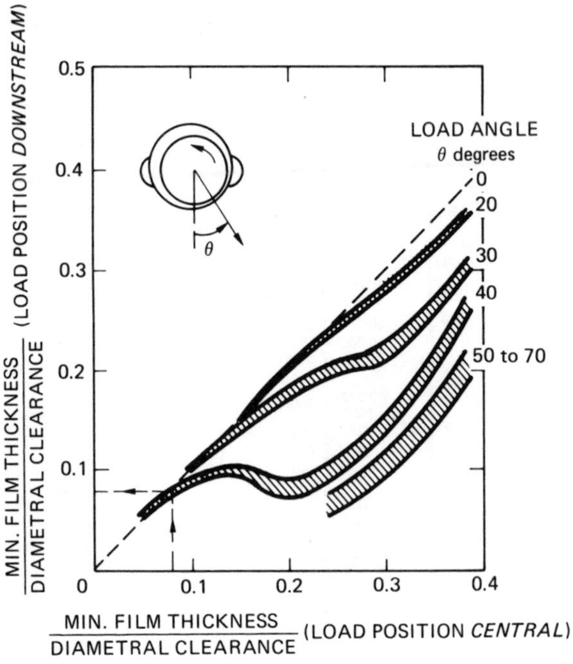

Fig. 9.11. Influence of load position on minimum film thickness (with load position downstream of centre)

EFFECT OF MISALIGNMENT ON FILM THICKNESS

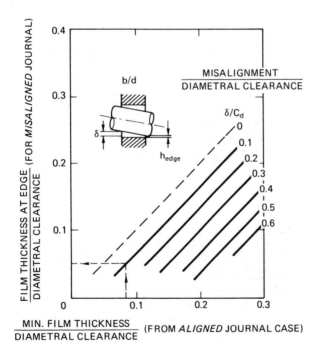

Fig. 9.12. Influence of misalignment on minimum film thickness

Bearings in high speed machines tend to have high power losses and oil film temperatures.

Machines with high speed rotating parts tend to be prone to vibration and this can be reduced by the use of bearings of an appropriate design.

Avoiding problems which can arise from high power losses and temperatures

Possible problem	Conditions under which it may occur	Possible solutions
Loss of operating clearance when starting the machine from cold	Designs in which the shaft may heat up and expand more rapidly than the bearing and its housing. Tubular shafts are prone to this problem	1. Avoid designs with features that can cause the problem 2. Design with diametral clearances towards the upper limit
	Housings of substantial wall thickness e.g. a housing outside diameter > 3 times the shaft diameter	3. Use a lubricant of lower viscosity if possible 4. Control the acceleration rate under cold starting conditions
	Housings with a substantial external flange member in line with the bearing	5. Preheat the oil system and machine prior to starting
Loss of operating clearance caused by the build up of corrosive deposits on a bearing or seal The deposit usually builds up preferentially at the highest temperature area, such as the position of minimum oil film thickness	Corrosion and deposition rates increase at higher operating temperatures A corrosive material to which the bearing material is sensitive needs to be present in the lubricating oil	1. Determine the chemical nature of the corrosion and eliminate the cause, which may be:- (a) an external contaminant mixing with the lubricant (b) an oil additive 2. Change the bearing material to one that is less affected by the particular corrosion mechanism 3. Attempt to reduce the operating temperature
Loss of operating clearance from the build up of deposits from microbiological contaminants The deposit usually builds up down-stream of the minimum film thickness where any water present in the lubricant tends to evaporate	The presence of condensation water in the lubricating oil and its build up in static pockets in the system Temperatures in low pressure regions of the oil films which exceed the boiling point of water	1. Modify the oil system to eliminate any static pockets, particularly in the oil tank 2. Occasional treatment of the lubrication system with biocides 3. Raise the oil system temperature if this is permissible
Increased operating temperatures arising from turbulence in the oil film	High surface speeds and clearances combined with low viscosity lubricants (see Fig. 10.1)	1. Check whether reduced bearing diameter or clearance may be acceptable 2. Accept the turbulence but check that the temperature rises are satisfactory
Increased operating temperatures arising from churning losses in the bearing housing	Thrust bearings are particularly prone to this problem because they are usually of a larger diameter than associated journal bearings	1. Keep the bearing housing fully drained of oil 2. In bearings with separate pads such as tilting pad thrust bearings, feed the pads by individual sprays

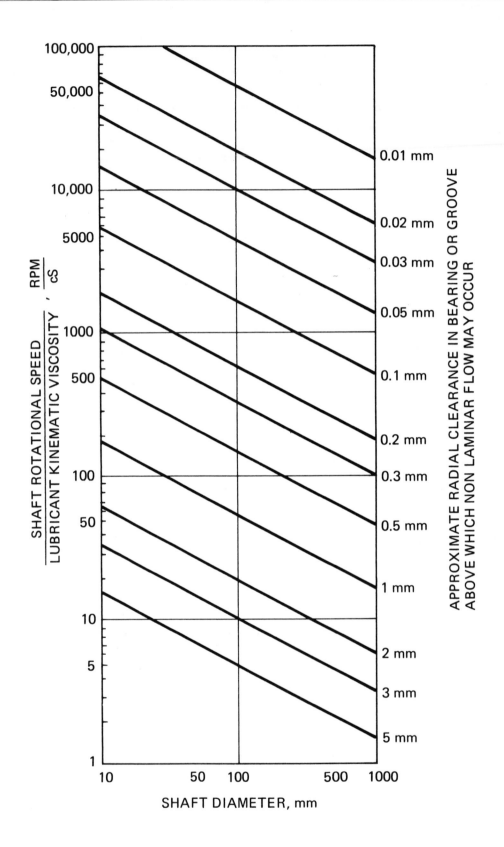

Fig. 10.1. Guidance on the occurrence of non laminar flow in journal bearings

Shaft lateral vibrations which may occur on machines with high speed rotors

Type of vibration	Cause of the vibration	Remarks
A vibration at the same frequency as the shaft rotation which tends to increase with speed	Unbalance of the rotor	Can be reduced by improving the dynamic balance of the rotor
A vibration at the same frequency as the shaft rotation which increases in amplitude around a particular speed	The rotor, as supported in the machine, is laterally flexible and has a natural lateral resonance or critical speed at which the vibration amplitude is a maximum	The response of the rotor in terms of vibration amplitude will depend on a balance between the damping in the system and the degree of rotor unbalance
A vibration with a frequency of just less than half the shaft rotational speed which occurs over a range of speeds	The rotor is supported in lightly loaded plain journal bearings which can generate half speed vibration (see Fig. 10.2). The actual frequency is generally just less than half shaft speed due to damping	An increase in the specific bearing loading by a reduction in bearing width can help. Alternatively bearings with special bore profiles can be used (see Fig. 10.3)
A vibration with a frequency of about half the shaft rotational speed, which shows a major increase in amplitude above a particular speed	The rotor is supported in lightly loaded plain journal bearings and is reaching a rotational speed, equal to twice its critical speed, when the major vibration increase occurs	This severe vibration arises from an interaction between the bearings and the rotor. The critical speed of the rotor resonates with half speed vibration of the bearings. Machines with plain journal bearings generally have a maximum safe operating speed of twice their first critical speed

The diagram shows the mechanism of operation of a plain journal bearing when supporting a steady load from the shaft.

The shaft rotational movement draws the viscous lubricant into the converging clearance and generates a film pressure to support the load

If a load is applied which rotates at half the shaft speed, the working of the bearing is not easy to visualise

OIL FEED IN THIS ZONE

REGION OF MINIMUM OIL FILM THICKNESS

RESULTANT PRESSURE GENERATION WHICH SUPPORTS THE LOAD, W

CONVERGING OIL FILM

AN EQUIVALENT DIAGRAM IS

Plain journal bearings cannot support loads which rotate at half the shaft speed.

Half speed loads arise if a bearing carries simultaneously a steady load and a load rotating at shaft speed, which are of equal magnitude

In this arrangement there is no resultant dragging of viscous lubricant into the loaded region, and no load carrying film is generated

If a bearing is lightly loaded the shaft tends to sit near the centre of the clearance space when operating. Any tendency then, to precess or vibrate at half the shaft rotational frequency, builds up in amplitude, because the bearing cannot provide a restoring force for loads/movements at this frequency.

Lightly loaded journal bearings tend therefore to generate shaft vibrations with a frequency of about half the shaft rotational speed

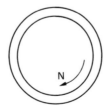

Fig. 10.2. The mechanism by which lightly loaded plain journal bearings tend to vibrate the shaft at about half its rotational speed

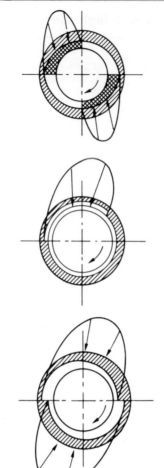

Lemon bore bearing
If the bearing is machined with shims between the joint faces, which are then removed for installation, the resultant bore is elliptical. When the shaft rotates, hydrodynamic clamping pressures are generated which restrain vibration

Dammed groove bearing
A wide and shallow central part circumferential groove in the upper half of the bearing terminates suddenly, and generates a hydrodynamic pressure which clamps downwards onto the shaft

Offset halves bearing
This can be made by machining the bore of two half bearing shells with a lateral offset and then rotating one shell about a vertical axis. This produces two strong converging oil films with high clamping pressures. This bearing, however, demands more oil flow than most other types. There are advantages in rotating the shells in the housing so that the pressure pattern that is generated aligns with the external load

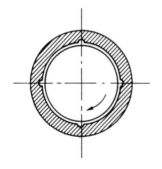

Multi-lobed bearing
A number of pads with a surface radius that is greater than that of the shaft are machined onto the bearing bore. This requires a broach or special boring machine. Each pad produces a converging hydro-dynamic film with a clamping pressure which stabilises the shaft

Preset of pads
For increased effectiveness the pads of multi lobe and tilting pad journal bearings need to be preset towards the shaft.

The typical presets commonly used are in the range of 0.6 to 0.8

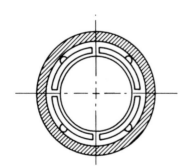

Tilting pad bearing
The shaft is supported by a number of separate pads able to pivot relative to an outer support housing. Each can generate stabilising hydrodynamic pressures

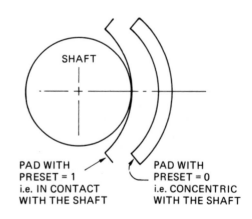

SHAFT

PAD WITH
PRESET = 1
i.e. IN CONTACT
WITH THE SHAFT

PAD WITH
PRESET = 0
i.e. CONCENTRIC
WITH THE SHAFT

Fig. 10.3. Bearings with special bore profiles to give improved shaft stability

ROTOR CRITICAL SPEEDS

The speed of a rotor at which a resonant lateral vibration occurs corresponds to the natural resonant frequency in bending of the rotor in its supports. This frequency corresponds closely to the ringing tone frequency which can be excited by hitting the rotor radially with a hammer, while it is sitting in its bearing supports. If the supports have different flexibilities in, for example, vertical and horizontal directions, such as may occur with floor mounted bearing pedestals, there will be two critical speeds.

The critical speed of a rotor can be reduced substantially by adding overhung masses such as drive flanges or flexible couplings at the ends of the rotor shaft. Figure 10.4 gives guidance on these effects.

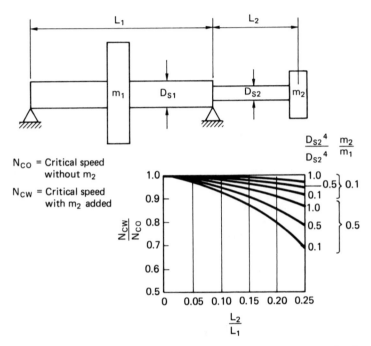

To determine the effect of additional mass at the shaft end, calculate $\dfrac{L_2}{L_1}$ and $\dfrac{D_{S2}{}^4}{D_{S1}{}^4}$

for the rotor, and also the mass ratio $\dfrac{m_2}{m_1}$, and read off an approximate value for

the critical speed reduction from the graph.

Fig. 10.4. The effect of an overhung mass such as a flexible coupling on the critical speed of a shaft

ROTORDYNAMIC EFFECTS

A full rotordynamic analysis of a machine tends to be complex largely because plain journal bearings give cross coupling effects. That is, a force applied by the shaft to the oil film, produces motion not only in line with the force, but also at right angles to it. This arises from the nature of the action of the hydrodynamic films in which the resultant pressure forces are not in line with the eccentric position of the shaft, within the bearing clearance.

Basic points of guidance for design can, however, be stated:

1 The most important performance aspect is the rotor response, in terms of its vibration amplitude.
2 The response is very dependent on the design of the journal bearings and the amount of damping that they can provide. Bearings with full oil films provide the most damping.
3 The likely mode shapes of the shaft need to be considered and the bearings should be positioned away from the expected position of any nodes. This is because at these positions the shaft has negligible radial movement when vibrating and bearings positioned at these nodes can therefore provide very little damping.

SELECTION OF PLAIN OR ROLLING BEARINGS

Characteristic	Rolling	Plain
Relative cost	High	Low
Weight	Heavier	Lighter
Space requirements:		
Length	Smaller	Larger
Diameter	Larger	Smaller
Shaft hardness	Unimportant	Important with harder bearings
Housing requirements	Usually not critical	Rigidity and clamping most important
Radial clearance	Smaller	Larger
Toleration of shaft deflections	Poor	Moderate
Toleration of dirt particles	Poor with hard particles	Moderate, depending on bearing hardness
Noise in operation	Tend to be noisy	Generally quiet
Running friction:		
Low speeds	Very low	Generally higher
High speeds	May be high	Moderate at usual crank speeds
Lubrication requirements	Easy to arrange. Flow small except at high speeds	Critically important pressure feed and large flow
Assembly on crankshaft	Virtually impossible except with very short or built-up crankshafts	Bearings usually split, and assembly no problem

At the present time the choice is almost invariably in favour of plain bearings, except in special cases such as very high-speed small engines, and particularly petroil two-strokes.

SELECTION OF TYPE OF PLAIN BEARING
Journal bearings

Direct lined	Insert liners
Accuracy depends on facilities and skill available	Precision components
Consistency of quality doubtful	Consistent quality
First cost may be lower	First cost may be higher
Repair difficult and costly	Repair by replacement easy
Liable to be weak in fatigue	Will generally sustain higher peak loads
Material generally limited to white metal	Range of available materials extensive

Thrust bearings

Flanged journal bearings	Separate thrust washer
Costly to manufacture	Much lower first cost
Replacement involves whole journal/thrust component	Easily replaced without moving journal bearing
Material of thrust face limited in larger sizes	Range of materials extensive
Aids assembly on a production line	Aligns itself with the housing

SELECTION OF PLAIN BEARING MATERIALS

Properties of typical steel-backed materials

Lining materials	Nominal composition %	Lining or overlay thickness		Relative fatigue strength	Guidance peak loading limits		Recommended Journal hardness V.P.N.
		mm	in		MN/m²	lbf/in²	
Tin-based white metal	Sn 87 Sb 9	Over 0.1	Over 0.004	1.0	12–14	1800–2000	160
	Cu 4 Pb 0.35 max.	Up to 0.1	Up to 0.004	1.3	14–17	2000–2500	160
Tin-based white metal with cadmium	Sn 89 Sb 7.5 Cu 3 Cd 1		No overlay	1.1	12–15	1800–2200	160
Sintered copper–lead, overlay plated with lead–tin	Cu 70	0.05	0.002	1.8	21–23	3000–3500(1)	230
	Pb 30	0.025	0.001	2.4	28–31	4000–4500(1)	280
Cast copper–lead, overlay plated with lead–tin or lead–indium	Cu 76 Pb 24	0.025	0.001	2.4	31	4500(1)	300
Sintered lead–bronze, overlay plated with lead–tin	Cu 74 Pb 22 Sn 4	0.025	0.001	2.4	28–31	4000–4500(1)	400
Aluminium–tin	Al 60 Sn 40	No overlay		1.8	21–23	3000–3500	230
Aluminium–tin	Al 80 Sn 20	No overlay		3	42	6000	230
Aluminium–tin (1) plated with lead–tin	Al 92 Sn 6 Cu 1 Ni 1	No overlay		3.5	48	7000	400
Aluminium–tin–silicon (2)	Al 82 Sn 12 Si 4 Cu 2	No overlay		3.7	52	7500	250

(1) Limit set by overlay fatigue in the case of medium/large diesel engines.
(2) Particularly suitable for use with nodular iron crankshafts.

Suggested limits are for big-end applications in medium/large diesel engines and are not to be applied to crossheads or to compressors. Maximum design loadings for main bearings will generally be 20% lower.

The fatigue strength of bearing metals varies with their thickness. As indicated in the previous table the fatigue strength of the overlay on a bearing material often provides a limit to the maximum load.

The graph shows the relative fatigue strength of two types of overlay material.

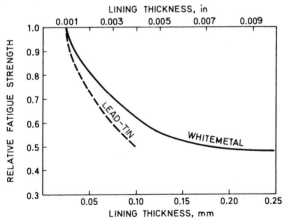

Automotive diesel and petrol engines

In smaller engines, thinner overlay plating can be used and the maximum loading is then no longer limited by fatigue of the overlay. With the exception of white metals the specific loading limits for smaller automotive diesel and petrol engines may by considerably higher than the values quoted in the table for medium/large diesel engines. Lead bronzes and tin–aluminium materials are available suitable for loads up to 7500 p.s.i. in these applications.

EFFECT OF FILM THICKNESS ON PERFORMANCE

Many well-designed bearings of modern engines tend to reach their limit of performance because of thin oil film conditions rather than by fatigue breakdown.

A vital factor determining the film thickness is the precise shape of the polar load diagram compared with the magnitude of the load vector. With experience it is possible to assess, from the diagram, whether conditions are likely to be critical and thus to determine whether computation of oil film thickness, peak pressures and power losses is advisable. The existence of load vectors rotating at half shaft speed is undesirable since they tend to reduce the oil film thickness.

Factors promoting thin oil films

Polar load diagrams, with features, are shown below.

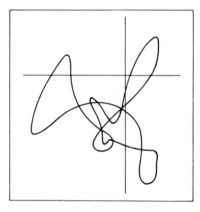

Two firing peaks combined with arc where load vector is travelling approximately at half shaft speed. (Diagram is typical of V-engine main bearing).

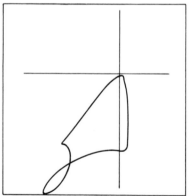

No half-speed vector but forces of large magnitude mostly directed in a limited quadrant. Journal dwells in one position in bearing during cycle, giving no chance for squeeze-film effects to assist.

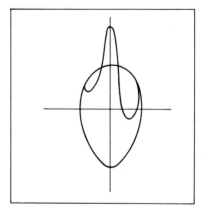

Big-end diagram with large inertia loop, promoting heavy loading on one area of crankpin throughout cycle. The resulting local overheating reduces the oil viscosity and gives thin oil films.

Computation of film thickness

If the film thickness conditions are likely to be critical, computation of the theoretical film thickness is desirable and several computer programmes are available.

Typical computer programme assumptions	Interpretation of results
Bearings are circular Journals are circular Perfect alignment Uniform viscosity around the bearing Effect of pressure on viscosity is ignored Surfaces are rigid Crankcase and crankshaft do not distort	The results cannot be used absolutely to determine whether any given bearing will or will not withstand the conditions of operation However, they are useful as a means of comparison with similar engines, or to indicate the effect of design changes. They provide a guided estimate of the probability of success

Depending on which programme is used, most bearings which are known to operate satisfactorily have computed minimum film thicknesses that are no lower than the following:

Mains   0.0025–0.0042 mm (0.0001–0.000 17 in)
Big ends 0.002–0.004 mm (0.000 08–0.000 15 in)

Means of improving oil film thickness

Modify crankshaft balance to reduce magnitude of rotating component of force.

Modify firing order, e.g. to eliminate successive firing of two or more cylinders adjacent to one bearing.

Increase bearing area to reduce specific loading, increase of land width being more effective than increase of diameter.

For big-end bearing, reduce reciprocative and/or rotating mass of con-rod.

Aim at acceptable compromise between factors including:

Firing loads
Inertia loads
Crankshaft stiffness
Torsional characteristics
Engine balance
Stiffness of big-end eye
Overall length of assembly

Each factor must be considered both on its own merits and in relation to others on which it has an effect.

BEARING FEATURES

General rules for grooving in crankshaft bearings

Never use a groove unless for a valid reason, and under no circumstances use a groove reaching in an axial direction.

For many engine applications a plain, central circumferential groove is used, e.g. to permit oil to be fed to a main bearing and thence, without interruption, to a big-end bearing via a drilled connecting rod to a small-end bearing.

Recommended groove configurations

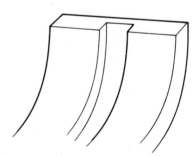

The groove may be formed within the bearing wall provided the wall thickness is adequate.

Undesirable grooving features

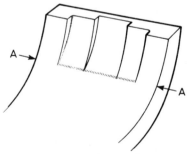

Gutterways or oil pockets at the horns. These do not trap dirt and are not needed for axial distribution of oil in a typical modern bearing. They can cause local overloading along line A–A, and fatigue breakdown.

Recommended groove configurations

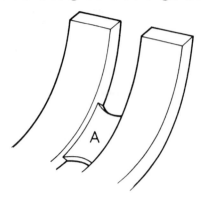

With thin shells the groove can take the form of a slot, communicating if necessary with a corresponding groove round the housing. The two halves of the shell are connected by bridge pieces as at *A*.

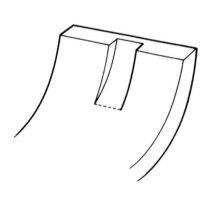

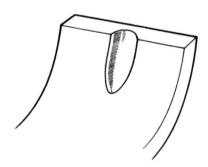

In order to leave a heavily loaded area of a bearing uninterrupted, a circumferential groove may be partial rather than continuous. The ends of such a groove must be blended gently into the plain surface to avoid sudden starting or stopping of the oil flow into or out of the groove and consequent erosion of the bearing.

Undesirable grooving features

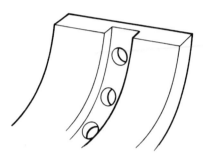

Rows of holes in oil groove, communicating with corresponding groove in bearing housing, can promote local recirculation of oil, leading to excessive temperatures.

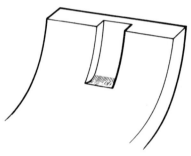

Abrupt endings to partial oil grooves can cause erosion of the surface downstream of the groove end.

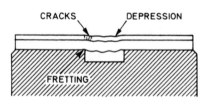

Any hole, slot or groove in the housing behind a bearing is dangerous because it promotes undesirable deflection, fretting and, in extreme cases, cracking of the bearing.

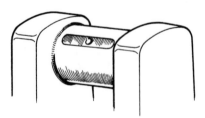

Any flat or depression in a journal surface, other than a blended hole in line with the bearing oil groove, is liable to cause severe local overheating.

Locating devices

Recommended

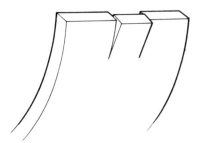

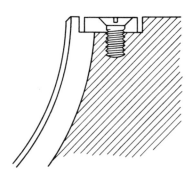

Nick, tang or lug locates shell in axial direction only. Lugs should be on one joint face of each shell with both lugs positioned at the same housing joint face.

Back face of nick must not foul slot or notch in housing, and nick should not be in heavily loaded area of bearing.

Bearing must be arranged with joint face in line with housing joint face.

Button stops—useful for large and heavy bearing shells —to retain upper halves in housings against falling out by own weight. Button may be any shape but is usually circular, and must be securely screwed to housing with small clearance all round and at upper face.

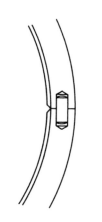

Dowel, tight fitting in one joint face, clearance in the other, locates shells to each other. More susceptible to damage during assembly, does not locate shells in housing and cannot be used on thin shells.

Not recommended

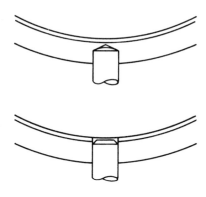

Dowel in crown—weakens shell, may induce fatigue or, if hole pierces shell, may cause erosion and oil film break-down. Will not prevent a seized bearing from rotating, and when this happens damage is aggravated.

Redundant locating devices result in 'fighting' between them and possible damage to bearings.

The correct conditions of installation are particularly important with dynamically loaded bearings.

LUBRICANT FEED SYSTEM

Except for small low-cost machines, oil feed is by a pressurised system consisting of a sump or reservoir, a mechanical pump with pressure-regulating valve and by-pass, one or more filters and an arrangement of pipes or ducts. The capacity of the system must be adequate to feed all bearings and other components even after maximum permissible wear has developed.

A guide to the flow through a conventional central circumferentially grooved bearing is given by

$$Q = \frac{kpC_d{}^3}{\eta} \cdot \frac{d}{b} (1.5\, \varepsilon^2 + 1)$$

where Q = flow rate, m³/sec (gal/min)

p = oil feed pressure, N/m² (lbf/in²)

C_d = diametral clearance, m (in)

η = dynamic viscosity, Ns/m² (cP)

d = bearing bore, m (in)

b = land width, m (in)

ε = eccentricity ratio

k = constant

= 0.0327 for SI units

(4.86×10^4 for Imperial units)

For most purposes it is sufficient to calculate the flow for a fully eccentric shaft, i.e. where $\varepsilon = 1$.

As a guide to modern practice, in medium/large diesel engines, oil-flow requirements at 3.5×10^5 N/m² (50 lbf/in²) pressure are as follows:

Bedplate gallery to mains (with piston cooling),
0.4 l/min/h.p. (5 gal/h/h.p.)
Mains to big end (with piston cooling),
0.27 l/min/h.p. (3.5 gal/h/h.p.)
Big ends to pistons (with oil cooling),
0.15 l/min/h.p. (2 gal/h/h.p.)
With uncooled pistons, total flow,
0.25 l/min/h.p. (3 gal/h/h.p.)

Velocity in ducts

On suction side of pump, 1.2 m/s (4ft/s).
On delivery side of pump, 1.8–3.0 m/s (6–10 ft/s).

Pressure

Modern high-duty engines will generally use delivery pressures in the range 2.8×10^5 to 4.2×10^5 N/m² (40–60 lbf/in²) but may be as high as 5.6×10^5 N/m² (80 lbf/in²).

Filtration

With the tendency to operate at very thin minimum oil films, filtration is specially important.

Acceptable criteria are that full-flow filters should remove:

100% of particles over 15 μm
95% of particles over 10 μm
90% of particles over 5 μm

Continuous bypass filtration of approximately 10% of the total flow may be used in addition.

INFLUENCE OF ENGINE COMPONENT DESIGN ON BEARING DESIGN AND PERFORMANCE

Housing tolerances

Geometric accuracy (circularity, parallelism, ovality) should be to H6 tolerances.

Lobing or waviness of the surface not to exceed 0.0001 of diameter.

Run-out of thrust faces not to exceed 0.0003 of diameter, total indicator reading.

Surface finish:

Journals 0.2–0.25 μm Ra (8–10 μin cla)
Gudgeon pins 0.1–0.16 μm Ra (4–6 μin cla)
Housing bores 0.75–1.6 μm Ra (30–60 μin cla)

Alignment of adjacent housing should be within 1 in 10 000 to 1 in 12 000.

Bearing housing bolts

These must be stressed to take with safety the sum of the loads due to compressing the bearing in its housing, and those due to the dynamic forces acting on the journal.

Housing stiffness

Local deflections under load have a disastrous effect on the fatigue strength of a bearing, causing fretting on the back, increased operating temperature and lower failure load. Housings, particularly connecting-rod eyes, should be as nearly as possible of uniform stiffness all round and the bolts must be positioned so as to minimise distortion of the bore. The effect of tightening bolts positioned too far from the bore is shown, exaggerated, below.

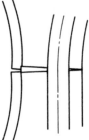

Crankshaft details

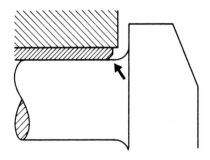

Normal fillet must clear chamfer on bearing under all conditions.

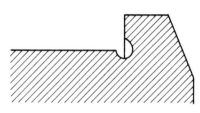

Undercut fillet, useful to help obtain maximum possible bearing width, and facilitates re-grinding.

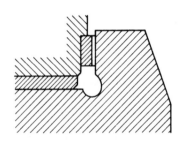

Thrust faces on crank cheeks must always be wider, radially, than soft surface of thrust ring or flange, to avoid throttling oil outlet when wear develops.

Grinding

Direction of grinding, especially on nodular iron shafts, is important and should be in the same direction as the bearing will move relative to the shaft.

Complete removal of friable or 'white' layer from nitrided journals is essential. This implies fine grinding or honing, approximately 0.025–0.05 mm (0.001–0.002 in) from surface.

SELECTION

Types of replaceable plain bearing

	Bush			Half bearing		
	SOLID	SPLIT	CLENCHED	THINWALL	MEDIUM WALL	THICK WALL
Normal housing	Solid	Solid	Solid	Split	Split	Split
Location feature	None	None	None	Nick	Nick or button stop	Dowel or button stop
Bearing material	Monometal*	Bimetal	Bimetal	Bimetal or trimetal	Monometal, bimetal or trimetal	Bimetal or trimetal
Manufacture	Cast or machine from solid	Wrap from flat strip	Wrap from flat strip	Press from flat strip	Cast lining on preformed blanks	Cast lining on preformed blanks
Forming of outside diameter	Machined	As pressed	Ground	As pressed	Fine-turned or ground	Fine turned
Other points to note when selecting a suitable type	Mainly below 50 mm dia.	Cheaper than solid in quantity	Higher precision	Cheapest for mass production	Moderate quantities 10–100 off	Made in pairs for 1–10 off

* Rarely large solid bimetal bushes are made for end assembly in solid housings. The wall thickness would approximate to that of medium-wall half bearings.

Double split bushes, with halves directly bolted together have been used in the past for turbine bearings, and some gearbox and ring oiled bearings. These are not now common in modern designs.

Double-flanged half bearings (thin and thick wall) and single flanged (solid) bushes are also used.

An oil distribution groove may be cut circumferentially in the outside diameter of thick-wall bearings only.

Choice of wall thickness

In the past, the shaft was chosen as the basis. The emphasis is now changing to standardising on the housing size.

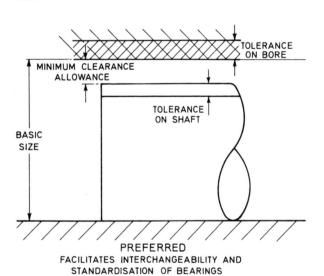

PREFERRED
FACILITATES INTERCHANGEABILITY AND
STANDARDISATION OF BEARINGS

NON PREFERRED
PRIMARY USE WHEN PREGROUND BAR
STOCK CAN BE USED FOR JOURNAL

Solid bushes and thick-wall bearings

Bore and outside diameter in metric sizes are chosen from preferred number series, the wall thickness varying accordingly but generally, within range shown in figures.

Wrapped bushes, thin- and medium-wall half bearings

The housing for metric sizes is based on preferred number series (ISO Recommendation R497 Series R40′) and wall thickness chosen from preferred thicknesses in table, but generally within range shown in figures.

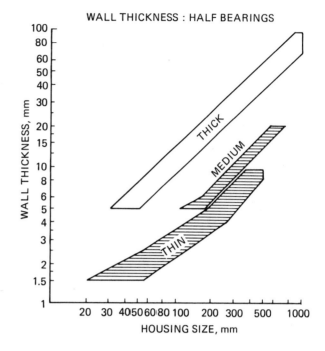

WALL THICKNESS : HALF BEARINGS

Preferred wall thickness (mm) and typical tolerances

Wrapped bushes		Half bearings			
		Thin wall		Medium wall	
Thickness	Tolerance	Thickness	Tolerance	Thickness	Tolerance
0.75*	0.050	1.5	0.008	4.0	0.020
1.0	0.050	1.75	0.008	5.0	0.020
1.5	0.050	2.0	0.008	6.0	0.020
2.0	0.080	2.5	0.010	8.0	0.025
2.5	0.100	3.0	0.015	10.0	0.025
3.0	0.100	3.5	0.015	12.5	0.030
3.5	0.100	4.0	0.020	15.0	0.030
		5.0	0.020	20.0	0.040
		6.0	0.025	25.0	0.040
		8.0	0.025		
		10.0	0.030		
		12.0	0.030		

* Non-preferred

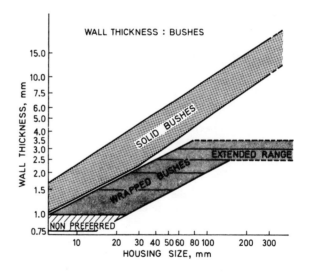

WALL THICKNESS : BUSHES

Wall thickness tolerance

Bushes are generally supplied with an allowance for finishing the bore *in situ* by boring, broaching, reeming, or ball burnishing, although prefinished bushes may be available to special orders.

Half bearings are generally machined to close tolerances on wall thickness and may be installed without further machining.

METHODS OF MEASUREMENT AND CHOICE OF HOUSING FITS

Ferrous-backed bearings in ferrous housings

Solid bushes

Outside diameter is measured.

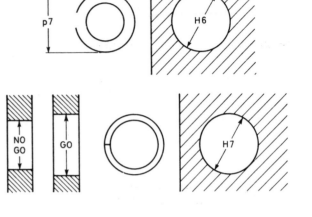

Split bushes

Outside diameter is measured in GO and NO GO ring gauges.

Typical ring gauge sizes (mm) are:

NO GO ring gauge
Below 30 ϕ Maximum housing $+0.035$
Above 30 ϕ Maximum housing $+0.008$ $\sqrt{D}$

GO ring gauge
Below 30 ϕ Maximum housing $+0.060$
Above 30 ϕ Maximum housing $+0.015$ $\sqrt{D}$

Where D = housing diameter (mm).
Split bushes may also be checked under load in a manner similar in principle to that used for thin-wall half bearings.

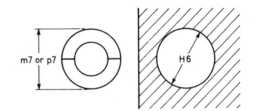

Thick-wall half bearings

Outside diameter is measured.
Split line not truly on centre line.
Bearings must be kept and stamped as pairs.

Thin-wall half bearing

Peripheral length of each half is measured under load W_c

$$W_c = 6000 \frac{b \times s}{D}$$

W_c = checking load (N)
b = width (mm)
s = wall thickness (mm)
D = housing diameter (mm)

'Nip' or 'Crush' is the amount by which the total peripheral length of both halves under no load exceeds the peripheral length of the housing. A typical minimum total nip n (mm) is given by: $n = 4.4 \times 10^{-5} \dfrac{D^2}{s}$ or 0.12 mm whichever is larger.

For bearings checked under load W_c

$$A = \frac{n}{2} - 0.050 \text{ mm}$$

The tolerance for A is grade 7 on peripheral length.

The nip value n gives a minimum radial contact pressure between bearing and housing of approximately 5 MN/m^2 for $D > 70$ mm increasing to 10 MN/m^2 at $D = 50$.

For white-metal lined bearings or similar lightly loaded bearings, minimum contact pressure and minimum nip values may be halved.

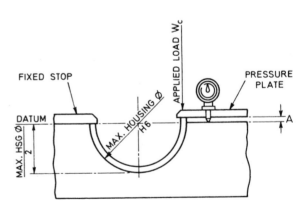

Non-ferrous backed bearings and housings

Allowance must be made in each individual case for the effects of differential thermal expansion. Higher interference fits may be necessary, and possible distortions will require checking by fitting tests.

Free spread

Thin- and medium-wall bearings are given a small amount of extra spread across the joint face to ensure that both halves assemble correctly and do not foul the shaft in the region of the joint when bolted up.

Free spread may be lost when bearing heats up, particularly for copper and aluminium based lining alloys. The loss depends upon method of forming bearing, and thickness of lining. Initial minimum free spread should exceed the likely loss if bearing is required to be reassembled.

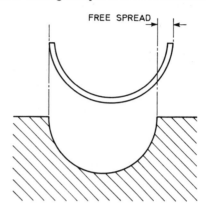

Typical minimum free spread (mm)

Outside diameter	Thin-wall aluminium or copper based lining	Medium-wall white-metal lining
50 ϕ	0.2–0.3	—
100 ϕ	1.0–2.0	0.2–0.5
300 ϕ	4.0–8.0	1.0–2.0

Note: these figures are for guidance only and specific advice should be obtained from the bearing manufacturers.

THE DESIGN OF THE HOUSING AND THE METHOD OF CLAMPING HALF BEARINGS

Locating the housing halves

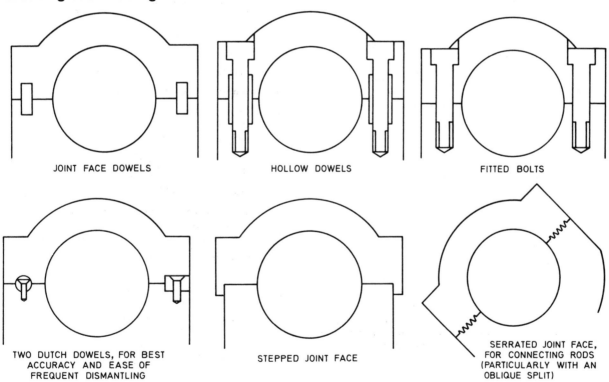

JOINT FACE DOWELS

HOLLOW DOWELS

FITTED BOLTS

TWO DUTCH DOWELS, FOR BEST ACCURACY AND EASE OF FREQUENT DISMANTLING

STEPPED JOINT FACE

SERRATED JOINT FACE, FOR CONNECTING RODS (PARTICULARLY WITH AN OBLIQUE SPLIT)

Positioning the bolts

These should be close to the back of the bearing.

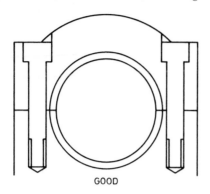

GOOD

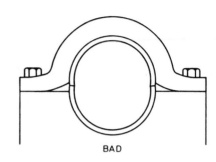

BAD

Choice of bolting loads

Load to compress nip

$$W = \frac{E\, b\, t\, m}{\pi\,(D-t)\times 10^6}$$

or

$$W = b\, t\, \phi \times 10^{-6}$$

whichever is smaller.

Extremely rigid housing (very rare)

Bolt load per side

$$W_b = 1.3\, W$$

(to allow for friction between bearing and housing).

Normal housing with bolts close to back of bearing

Bolt load per side

$$W_b \simeq 2\, W$$

(to allow for friction, and relative moments of bearing shell and line of bolt about outer edge of housing).

Note: If journal loads react into housing cap

$$W_b = 2\, W + \tfrac{1}{2}\, W_j$$

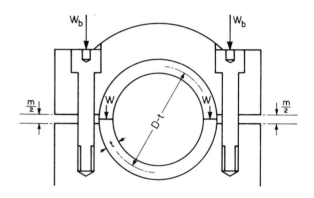

D = housing diameter (mm)

t = steel thickness $+\tfrac{1}{2}$ lining thickness (mm)

W = compression load on each bearing joint face (N)

m = sum of maximum circumferential nip on both halves (mm) [$m = n + \pi$(housing diametral tolerance) $+$ tolerance on nip]

E = elastic modulus of backing (N/m^2) (For steel $E = 0.21 \times 10^{12}$ N/m^2)

b = bearing axial width (mm)

W_b = bolt load required on each side of bearing to compress nip (N)

W_j = Maximum journal load to be carried by cap (N)

ϕ = yield stress of steel backing (N/m^2)

ϕ = varies with manufacturing process and lining, e.g.
white-metal lined bearing
$\phi = 350 \times 10^6$ N/m^2
copper based lined bearing
$\phi = 300\text{--}400 \times 10^6$ N/m^2
aluminium based lined bearing
$\phi = 600 \times 10^6$ N/m^2

Housing material

For non-ferrous housings, or for non-ferrous backed bearings in ferrous housings, allowance must be made for the effects of thermal expansion causing loss of interference fit. Specific advice should be obtained from bearing manufacturers.

Dynamic loads

Dynamic loads may result in flexing of the housing if this is not designed to have sufficient rigidity. Housing flexure can result in fretting between bearing back and housing, and in severe cases, in fatigue of the housing. Increase of interference fit with a corresponding increase in bolting load, may give some alleviation of fretting but stiffening the housing is generally more effective. Poor bore contour and other machining errors of the housing can lead to fretting or loss of clearance. Surface finish of housing bore should be 1 to 1.5 μm cla. Oil distribution grooves in the bore of the housing should be avoided for all except thick-wall half bearings.

Rotating loads

High rotating loads (in excess of 4–6 MN/m^2) such as occur in the planet wheels of epicyclic gearboxes may result in bushes very slowly creeping round in their housings and working out of the end. Location features such as dowels are unable to restrain the bush and it is normal practice in these circumstances to cast the lining material on the pin.

PRECAUTIONS WHEN FITTING PLAIN BEARINGS

Half bearings

(1) *Ensure cleanliness:* between bearing and housing, between bearing bore and joint face of bearing and housing, and in oil-ways and oil system.
(2) *Check free spread:* bearing should be in contact with housing in region of joint face.
(3) *Check nip:* tighten bolts, slacken one side to hand tight, use feeler gauge on housing joint. Tighten bolts and repeat on other side.
(4) *Tighten bolts:* to specified stretch or torque.
(5) *Check clearance:* between journal and assembled bore with leads or similar, particularly checking for distortions (local loss of clearance) near oil grooves or bearing joint face. (For new design of housing conduct fitting test: measure bore diameter with shaft removed.)
(6) *Oil surface:* before turning shaft.

Bushes

(1) *For insertion:* use oil or other suitable lubricant to facilitate pressing-in and prevent scoring of bush outside diameter. Use a 15° lead in chamfer in housing, and press in square.
(2) *Check assembled bore size:* with comparator or plug gauge.
(3) *Remove burrs and sharp edges from shaft:* to avoid damage to bush bore during assembly.
(4) *Check cleanliness:* in bush bore, at bush ends and in lubricant supply drillings.

SPHERICAL BEARINGS

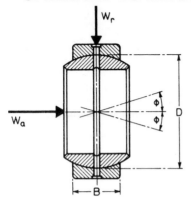

EQUIVALENT BEARING PRESSURE

$$p = \frac{W_r^2 + 6W_a^2}{W_r\, BD}$$

PROVIDED THAT

$$W_a < W_r$$

AS A RULE: $\phi < 8°$

Bearing life calculations:

$$L = f \cdot \left(\frac{p_o}{p}\right)^3 \times 10^5$$

L = bearing life, i.e. average number of oscillations to failure assuming unidirectional loading

f = life-increasing factor depending on periodical re-lubrication (ref. table below)

p_o = maximum bearing pressure if an average bearing-life of 10^5 number of oscillations is to be expected (ref. table below)

p = equivalent bearing pressure (ref. formula at left)

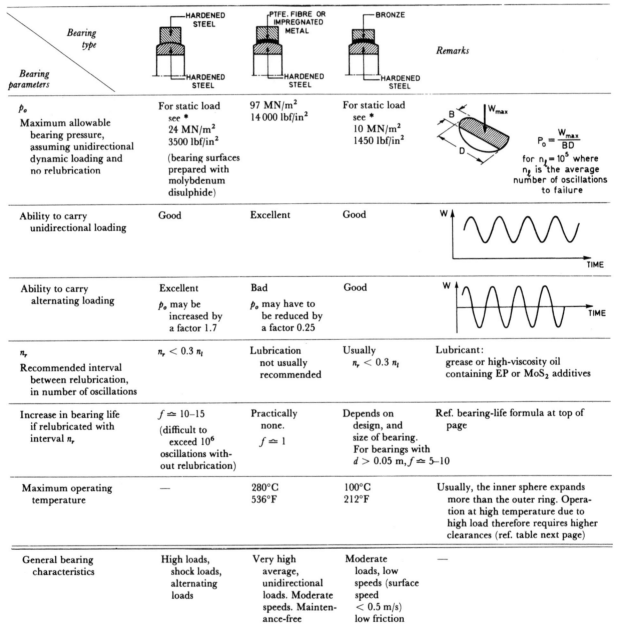

Bearing parameters \ Bearing type	HARDENED STEEL / HARDENED STEEL	PTFE, FIBRE OR IMPREGNATED METAL / HARDENED STEEL	BRONZE / HARDENED STEEL	Remarks
p_o Maximum allowable bearing pressure, assuming unidirectional dynamic loading and no relubrication	For static load see * 24 MN/m² 3500 lbf/in² (bearing surfaces prepared with molybdenum disulphide)	97 MN/m² 14 000 lbf/in²	For static load see * 10 MN/m² 1450 lbf/in²	$P_o = \dfrac{W_{max}}{BD}$ for $n_l = 10^5$ where n_l is the average number of oscillations to failure
Ability to carry unidirectional loading	Good	Excellent	Good	
Ability to carry alternating loading	Excellent p_o may be increased by a factor 1.7	Bad p_o may have to be reduced by a factor 0.25	Good	
n_r Recommended interval between relubrication, in number of oscillations	$n_r < 0.3\, n_l$	Lubrication not usually recommended	Usually $n_r < 0.3\, n_l$	Lubricant: grease or high-viscosity oil containing EP or MoS₂ additives
Increase in bearing life if relubricated with interval n_r	$f \simeq 10$–15 (difficult to exceed 10^6 oscillations without relubrication)	Practically none. $f \simeq 1$	Depends on design, and size of bearing. For bearings with $d > 0.05$ m, $f \simeq 5$–10	Ref. bearing-life formula at top of page
Maximum operating temperature	—	280°C 536°F	100°C 212°F	Usually, the inner sphere expands more than the outer ring. Operation at high temperature due to high load therefore requires higher clearances (ref. table next page)
General bearing characteristics	High loads, shock loads, alternating loads	Very high average, unidirectional loads. Moderate speeds. Maintenance-free	Moderate loads, low speeds (surface speed < 0.5 m/s) low friction	—

* The table is based on dynamic load conditions. For static load conditions, where the load-carrying capacity of the bearing is based on bearing-surface permanent deformation, not fatigue, the load capacity of steel bearings may reach $10 \times p_o$ and of aluminium bronze $5 \times p_o$.

Spherical bearings are commonly used where the oscillatory motion is a result of a misalignment, which may be intentional or not. A lateral oscillatory motion is then often combined with a rotatory motion about the bearing axis. If the frequency of rotation is high, it may require a separate bearing fitted inside the spherical assembly.

Clearance

Bore		Radial clearance C_r					
D		C2		Normal		C3	
From	Up to	Min.	Max.	Min.	Max.	Min.	Max.
mm		μm		μm		μm	
—	12	8	32	32	68	68	104
12	20	10	40	40	82	82	124
20	35	12	50	50	100	100	150
35	60	15	60	60	120	120	180
69	90	18	72	72	142	142	212
90	140	18	85	85	165	165	245
140	240	18	100	100	192	192	284
240	300	18	110	110	214	214	318

Clearance definitions :-

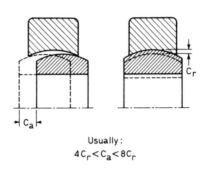

Usually :
$$4C_r < C_a < 8C_r$$

C2 CLEARANCE may somtimes be chosen if the loading is alternating

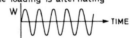

C3 CLEARANCE is sometimes required if the temperature difference between inner sphere and outer ring is large, or the bearing works in dirty conditions

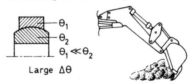

Large $\Delta\theta$

Fit

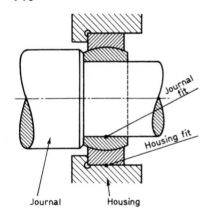

Contact point	Most common fit			Housing made of light alloy	Push fit in housing	Push fit on journal
	C2	Normal	C3			
Recommended journal fit	h6	m6	m6	—	m6	h6 Journal should be hardened
Recommended housing fit	J7	K7 or M7	M7	N7*	H7	M7

* In general, if the housing is made of light-metal alloy or other soft material, or the housing is thin-walled, a tighter fit is required. Recommended: one degree tighter fit than otherwise would have been chosen, i.e. M is replaced by N, J by K, etc.

ROLLING BEARINGS

Rolling bearings (i.e. ball, roller and needle bearing) may also be used for oscillatory motion, but preferably where the load is unidirectional, or at least varies with moderate gradient.

Static (or near static) loading	*Dynamic loading*

Static (or near static) loading

Bearings are selected on the basis of their load-carrying capacity.

Examples:

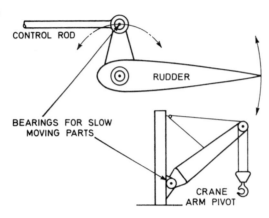

CONTROL ROD

RUDDER

BEARINGS FOR SLOW MOVING PARTS

CRANE ARM PIVOT

Rule:

Use manufacturers figures for the static load coefficient C_o multiplied by a factor f such that:

$f = 0.5$ for sensitive equipment (weights, recorders, etc.)
$f = 1.0$ for crane arms, etc.
$f = 5.0$ for emergency cases on control rods (e.g. for aircraft controls).

Dynamic loading

Bearings are selected on the basis of their required life before failure.

Examples:
small-end bearings in engines and compressors
plunger-pin bearings in crank-operated presses
connecting rods in textile machinery
connecting rods in wood reciprocating saws

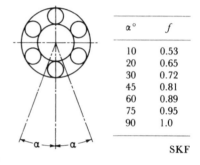

$\alpha°$	f
10	0.53
20	0.65
30	0.72
45	0.81
60	0.89
75	0.95
90	1.0

SKF

Rule:

(1) If $\alpha \geqslant 90°$, each oscillation is considered as a complete revolution, and the bearing life is determined as if the bearing was rotating.
(2) If $\alpha < 90°$, the equivalent load on the bearing is reduced by a factor f, taken from the table above. The calculations are then carried out as under (1).

ENGINE SMALL-END BEARINGS

Type of engine	2-Stroke		4-Stroke	
	Small and medium engines	Large engines		
Type of bearing and oil grooving	NEEDLE BEARINGS: single	A series of axial oil grooves interconnected by a circumferential groove	PLAIN BEARING	Single central circumferential oil groove. For small bearings, sometimes only oil hole(s)
Type of gudgeon pin	double	$d \approx B$ $d \approx D/3$		Surface finish better than 0.05 µm CLA Material: Surface hardened steel
Bearing material and allowable pressure*	21–35 MN/m² 3000–5000 lbf/in². The load capacity varies, refer to manufacturer's specifications	Phosphor bronze 25–30 MN/m² 3500–4300 lbf/in²	Lead bronze up to 25 MN/m² 3500 lbf/in²	Phosphor bronze up to 50 MN/m² 7100 Ibf/in²
Type of friction	Rolling friction	Mixed to boundary friction	Mostly mixed lubrication, but may be fully hydrodynamic under favourable condition	
Diametral clearance	Clearances not applicable. Fits: Pin/Piston *J6*; Ring/conrod *P7*	$\simeq 1$ µm/mm of d	1–1.5 µm/mm of d	$\simeq 1$ µm/mm of d
Remarks	Needle bearings are best suited where the loading is uni-directional	This type of bush is often made floating in a fixed steel bush	Liable to corrosion in plain mineral oil. An overlay of lead-base white metal will reduce scoring risk	

* Bearing pressure is based on projected bearing area, i.e. $B \times d$ (ref. sketch above).

ENGINE CROSSHEAD BEARINGS

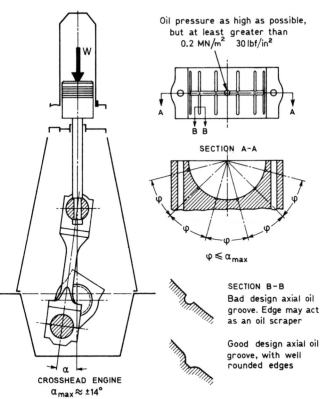

Oil pressure as high as possible, but at least greater than 0.2 MN/m² 30 lbf/in²

SECTION A-A

$\varphi \leq \alpha_{max}$

SECTION B-B
Bad design axial oil groove. Edge may act as an oil scraper

Good design axial oil groove, with well rounded edges

CROSSHEAD ENGINE
$\alpha_{max} \approx \pm 14°$

EXAMPLE OF OIL GROOVE DESIGN IN A CROSSHEAD BEARING OF CA 300mm DIAMETER. ALL MEASUREMENTS ARE IN MILLIMETRES (SECTION B–B)

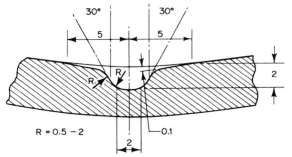

R = 0.5 – 2

Bearing materials *	Maximum allowable peak pressure	Diametral clearance (in μm/ mm of pin dia.)	Remarks
White metal (tin base)	7 MN/m² 1000 lbf/in²	Δ_1 0.5–0.7	Excellent resistance against scoring. Corrosion resistant. Low fatigue strength
Copper–lead	14 MN/m² 2000 lbf/in²	$2\Delta_2$ $\simeq 1$	High-strength bearing metal sensitive to local high pressure. Liable to corrosion by acidic oil unless an overlay of lead–tin or lead–indium is used ($\simeq 25\ \mu m$)
Tri-metal e.g. steel copper–lead white metal (lead)	14 MN/m² 2000 lbf/in²	$2\Delta_2$ $\simeq 1$	Same as above, but better resistance to corrosion, wear and scoring. Installed as bearing shells, precision machined

* Tin–aluminium is also a possible alloy for crosshead bearings, and spherical roller bearings have been used experimentally.

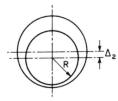

Old (and still common) practice:
Bearing metal scraped to conformity with wristpin over an arc of $\simeq 120$–150°. Mostly used for large, two-stroke marine diesel engines. Works mainly with boundary friction.

New practice:
Bearing precision machined to an exact cylinder with radius slightly greater than wristpin radius. Mainly hydrodynamic lubrication. In common use for 4-stroke engines, and becoming common on 2-stroke engines.

Local high pressures and thermal instability

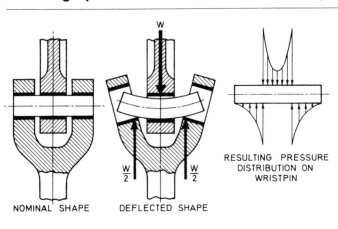

NOMINAL SHAPE   DEFLECTED SHAPE

RESULTING PRESSURE DISTRIBUTION ON WRISTPIN

Oscillating bearings in general, and crosshead bearings in particular, have a tendency to become thermally unstable at a certain load level. It is therefore of great importance to avoid local high pressures due to wear, misalignments or deflections such as shown in the figure at left.

In critical machine components, such as crosshead bearings, temperature warning equipment should be installed

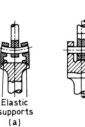

Central bearing

Elastic supports
(a)

SECTION A-A
(b)

Two possible solutions used in crosshead bearings:
(a) Elastic bearing supports
(b) Upper end of connecting rod acts as a partial bearing (central loading)

Squeeze action

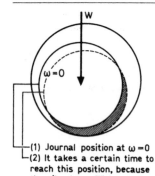

In spite of the fact that the angular velocity of the journal is zero twice per cycle in oscillating bearings, such bearings may still work hydrodynamically. This is due to the squeeze action, shown in the diagram. This squeeze action plays an important role in oscillating bearings, by preventing excessive metallic contact

(1) Journal position at $\omega = 0$
(2) It takes a certain time to reach this position, because the oil volume ▨ has to be 'squeezed' away. Before this occurs, $\omega > 0$

THE LOAD REVERSES ITS DIRECTION DURING THE CYCLE

On some bearings, the load reverses direction during the cycle. This will help to create a thicker oil film at velocity reversal, and thus the squeeze action will be more effective. Load reversal takes place in piston pin and crosshead bearings in 4-stroke engines, but not in 2-stroke engines. The latter engines are therefore more liable to crosshead bearing failure than the former

OSCILLATORY BEARINGS WITH SMALL RUBBING VELOCITY

In oscillatory bearings with small rubbing velocities, it is necessary to have axial oil grooves in the loaded zone, particularly if the load is unidirectional.

EXAMPLES OF GROOVE PATTERN

Oil lubricated bearing

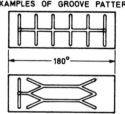

180°

Grease lubricated bearing

Grease lubricated bearing

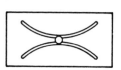

Bronze is a common material in oscillatory journal bearings with small rubbing velocity and large, unidirectional loading.

Bearings are often made in the form of precision machined bushings, which may be floating.

W up to 60 MN/m²; 8500 lbf/in².

$\Delta \simeq 1.2\ \mu m/mm$ of D.

ENGINE ROCKER BEARING

Usually bronze bush eg. Cu Sn8 (91-92% Cu)

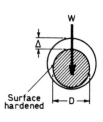

Surface hardened

Example of floating bush

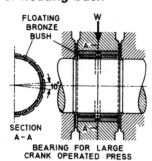

FLOATING BRONZE BUSH

SECTION A-A

BEARING FOR LARGE CRANK OPERATED PRESS

In the example left the projected bearing area is 0.018m² (278 in²) and the bearing carries a load of 2 MN ($\simeq$ 450.000 lbf).

$$W = 2\ MN\ (\simeq 450\,000\ lbf)$$

The bush has axial grooves on inside *and* outside. On the outside there is a circumferential groove which interconnects the outer axial grooves and is connected to the inner axial grooves by radial drillings.

SPHERICAL BEARINGS FOR OSCILLATORY MOVEMENTS (BALL JOINTS)

Types of ball joints

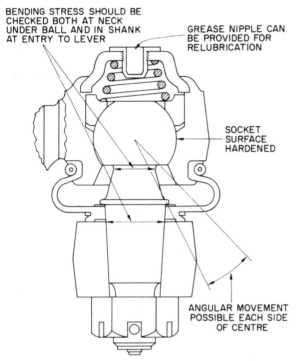

BENDING STRESS SHOULD BE CHECKED BOTH AT NECK UNDER BALL AND IN SHANK AT ENTRY TO LEVER

GREASE NIPPLE CAN BE PROVIDED FOR RELUBRICATION

SOCKET SURFACE HARDENED

ANGULAR MOVEMENT POSSIBLE EACH SIDE OF CENTRE

Fig. 14.1. Transverse type ball joint with metal surfaces (*courtesy:* Automotive Products Co. Ltd)

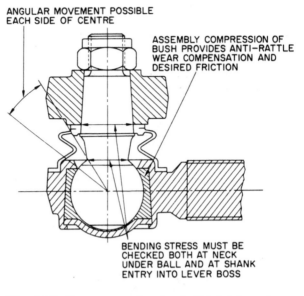

ANGULAR MOVEMENT POSSIBLE EACH SIDE OF CENTRE

ASSEMBLY COMPRESSION OF BUSH PROVIDES ANTI-RATTLE WEAR COMPENSATION AND DESIRED FRICTION

BENDING STRESS MUST BE CHECKED BOTH AT NECK UNDER BALL AND AT SHANK ENTRY INTO LEVER BOSS

Fig. 14.2. Transverse steering ball joint (*courtesy:* Cam Gears Ltd)

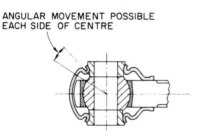

ANGULAR MOVEMENT POSSIBLE EACH SIDE OF CENTRE

Fig. 14.3. Straddle type joint shown with gaiters and associated distance pieces (*courtesy:* Rose Bearings Ltd)

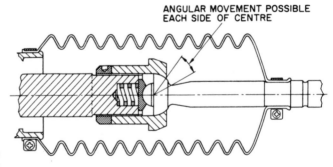

ANGULAR MOVEMENT POSSIBLE EACH SIDE OF CENTRE

Fig. 14.4. Axial ball joint (*courtesy:* Cam Gears Ltd)

Selection of ball joints

The many different forms of ball joints developed for a variety of purposes can be divided into two main types, *straddle mounted* [rod ends], and *overhung*. They may be loaded perpendicularly to, or in line with the securing axis. Working loads on ball joints depend upon the application, the working pressures appropriate to the application, the materials of the contacting surfaces and their lubrication, the area factor of the joint and its size. The area factor, which is the projected area of the tropical belt of width L divided by the area of the circle of diameter D, depends upon the ratio L/D. The relationship is shown in the graph (Fig. 14.6).

Transverse types are seldom symmetrical and probably have a near equatorial gap (Fig. 14.1) but their area factors can be arrived at from Fig. 14.6 by addition and subtraction or by calculation. For straddle and transverse type joints, either the area factor or an actual or equivalent L/D ratio could be used to arrive at permissible loadings, but when axially loaded joints are involved it is more convenient to use the area factor throughout and Fig. 14.6 also shows the area factor–L/D ratio relationship for axially loaded joints.

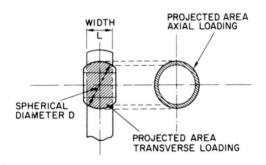

Fig. 14.5. Ball joint parameters

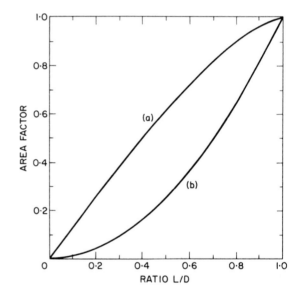

Fig. 14.6. Area factors (a) transverse and straddle type ball joints (b) axial type ball joints

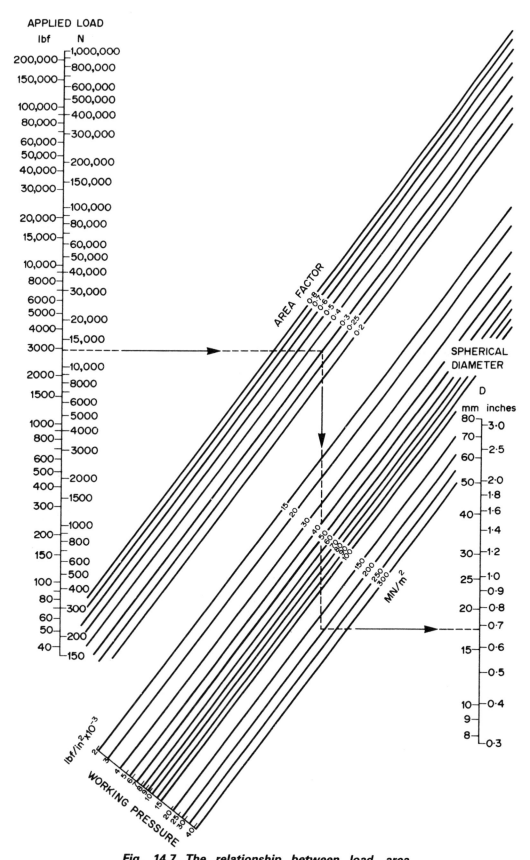

Fig. 14.7. The relationship between load, area factor, working pressure and spherical diameter of ball joints

A guide to the selection and performance of ball joints

Type	Straddle or rod end	Axial	Transverse
Angle	$\pm 10°$ to $\pm 15°$ with minimum shoulder on central pin, $\pm 30°$ to $40°$ with no shoulder on central pin	$\pm 25°$ to $\pm 30°$	$\pm 10°$ to $\pm 15°$ low angle $\pm 25°$ to $\pm 30°$ high angle
Main use	Linkages and mechanisms	Steering rack end connections	Steering linkage connections, suspension and steering articulations
Lubrication	Grease	Grease	Lithium base grease on assembly. Largest sizes may have provision for relubrication
Enclosure and protection	Often exposed and resistant to liquids and gases. Rubber gaiters available (Fig. 14.3)	Rubber or plastic bellows, or boot	Rubber or plastic seals, or bellows
Materials	*Inner.* Case or through hardened steel, hardened stainless steel, hardened sintered iron; possibly chromium plated *Outer bearing surfaces.* Aluminium bronze, naval bronze, hardened steel, stainless steel, sintered bronze, reinforced PTFE	*Ball.* Case-hardened steel *Bushes.* Case or surface hardened steel, bronze, plastic or woven impregnated	*Ball.* Case-hardened steel *Bushes.* Case or surface hardened steel, bronze, plastic or woven impregnated
Working pressures	Limiting static from 140 MN/m² to 280 MN/m² on projected area depending on materials. Wear limited on basis of 50×10^3 cycles of ± 25 at 10 cycles/min from 80MN/m² to 180 MN/m² depending on materials	35 to 50 MN/m² on maximum or measured forces	Approximately 15 MN/m² on plastic, 20 MN/m² on metal surface projected areas. Bending stress in the neck or shank which averages 15 times the bearing pressure limits working load. Fatigue life must also be considered
Area factors	0.42 to 0.64 with radial loads and 0.12 to 0.28 with axial loads	0.25	0.55 large angles 0.7 small angles
Remarks	No provision to take up wear which probably determines useful life. Use Fig. 14.7 for selection or consult manufacturer	Spring loaded to minimise rattle and play	Steering and suspension joints spring loaded to minimise rattle and play and to provide friction torque. Some plastic bush joints rely on compression assembly for anti-rattle and wear compensation

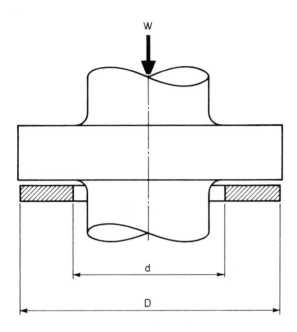

Plain thrust washers are simple and occupy little axial space. Their performance cannot be predicted with accuracy because their operation depends upon small-scale surface undulations and small dimensional changes arising from thermal expansion whilst running.

Thrust washers with radial grooves (to encourage hydrodynamic action) are suitable for light loads up to 0.5 MN/m² (75 lbf/in²), provided the mean runner speed is not less than the minimum recommended below according to lubricant viscosity.

Minimum sliding speeds to achieve quoted load capacity

Viscosity grade ISO 3448	Minimum sliding speed = $\pi n d_m$	
	m/s	in/s
100	2.5	100
68	4	160
46	6	240
32	8	320

Suitable materials

0.5 mm (approx) white metal on a steel backing, overall thickness 2–5 mm, with a Mild Steel Collar.

or

Lead bronze washer with a hardened steel collar.

Recommended surface finish for both combinations
Bearing 0.2–0.8 μm Ra
  (8–32 μin cla)
Collar 0.1–0.4 μm Ra
  (4–16 μin cla)

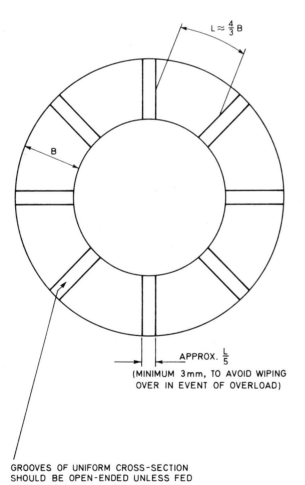

$L \approx \frac{4}{3} B$

B

APPROX. $\frac{L}{5}$
(MINIMUM 3mm, TO AVOID WIPING OVER IN EVENT OF OVERLOAD)

GROOVES OF UNIFORM CROSS-SECTION SHOULD BE OPEN-ENDED UNLESS FED WITH LUBRICANT AT HIGH PRESSURE

Estimation of approximate performance

Recommended maximum load:

$$W = K_1(D^2 - d^2)$$

Approximate power loss in bearing:

$$H = K_2 \, n \, d_m \, W$$

Lubricant flow rate to limit lubricant (oil) temperature rise to 20°C:

$$Q = K_3 \, H$$

Symbol and meaning	SI units	Imperial units
W load	N	lbf
H power loss	W	h.p.
Q flow rate	m^3/s	g.p.m.
n rotational speed	rev/s	r.p.s.
d, D	mm	in
$d_m \; (D + d)/2$	mm	in
K_1	0.3	48
K_2	70×10^{-6}	11×10^{-6}
K_3	0.03×10^{-6}	0.3

Lubricant feeding

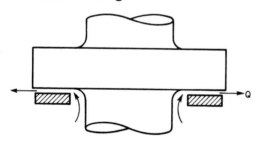

Lubricant should be fed to, or given access to, inner diameter of the bearing so that flow is outward along the grooves.

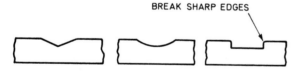

Suitable groove profiles

For horizontal-shaft bearings the grooves may have to be shallow (0.1 mm) to prevent excess drainage through the lower grooves, which would result in starvation of the upper pads. For bearings operating within a flooded housing a groove depth of 1 mm is suitable.

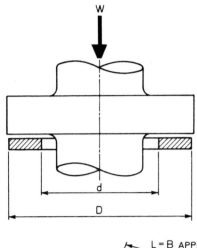

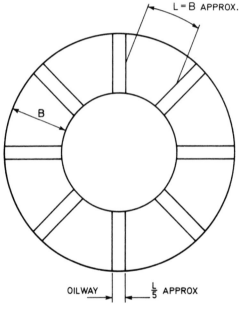

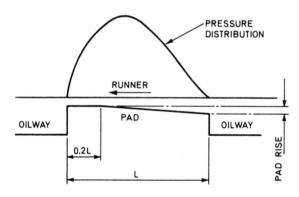

PROFILE ALONG PAD — UNI-DIRECTIONAL

Fig. 16.1 Bearing and pad geometry

BEARING TYPE AND DESCRIPTION

The bearing comprises a ring of sector-shaped pads. Each pad is profiled so as to provide a convergent lubricant film which is necessary for the hydrodynamic generation of pressure within the film. Lubricant access to feed the pads is provided by oil-ways which separate the individual pads. Rotation of the thrust runner in the direction of decreasing film thickness establishes the load-carrying film. For bi-directional operation a convergent–divergent profile must be used (see later). The geometrical arrangement is shown in Fig. 16.1

FILM THICKNESS AND PAD PROFILE

In order to achieve useful load capacity the film thickness has to be small and is usually in the range 0.005 mm (0.000 2 in) for small bearings to 0.05 mm (0.002 in) for large bearings. For optimum operation the pad rise should be of the same order of magnitude. Guidance on suitable values of pad rise is given in Table 16.1.

The exact form of the pad surface profile is not especially important. However, a flat land at the end of the tapered section is necessary to avoid excessive local contact stress under start-up conditions. The land should extend across the entire radial width of the pad and should occupy about 15–20% of pad circumferential length.

Table 16.1 Guidance on suitable values of pad rise

Bearing inner diameter d		Pad rise	
mm	inch	mm	inch
25	1	0.015–0.025	0.0006–0.001
50	2	0.025–0.04	0.001 –0.0016
75	3	0.038–0.06	0.0015–0.0025
100	4	0.05 –0.08	0.002 –0.0032
150	6	0.075–0.12	0.003 –0.0048
200	8	0.10 –0.16	0.004 –0.0064
250	10	0.12 –0.20	0.005 –0.008

It is important that the lands of all pads should lie in the same plane to within close tolerances; departure by more than 10% of pad rise will significantly affect performance (high pads will overheat, low pads will carry little load). Good alignment of bearing and runner to the axis of runner rotation (to within 1 in 10^4) is necessary. Poorly aligned bearings are prone to failure by overheating of individual pads.

GUIDE TO BEARING DESIGN

Bearing inner diameter should be chosen to provide adequate clearance at the shaft for oil feeding and to be clear of any fillet radius at junction of shaft and runner.

Bearing outer diameter will be determined, according to the load to be supported, as subsequently described. Bearing power loss is very sensitive to outer diameter, and conservative design with an unnecessarily large outside diameter should therefore be avoided.

Oil-ways should occupy about 15–20% of bearing circumference. The remaining bearing area should be divided up by the oil-ways to form pads which are approximately 'square'. The resulting number of pads depends upon the outer/inner diameter ratio—guidance on number of pads is given in Fig. 16.4

Safe working load capacity

Bearing load capacity is limited at low speed by allowable film thickness and at high speed by permissible operating temperature.

Guidance on safe working load capacity is given for the following typical operating conditions:

    Lubricant (oil) feed temperature, 50°C
    Lubricant temperature rise through bearing housing, 20°C

in terms of a basic load capacity W_b for an arbitrary diameter ratio and lubricant viscosity (Fig. 16.2), a viscosity factor (Fig. 16.3) and a diameter ratio factor (Fig. 16.4). That is:

Safe working load capacity $= W_b \times$ (Viscosity factor)
$\times$ (Diameter ratio factor)

Example

To find the necessary outer diameter D for a bearing of inner diameter $d = 100$ mm to provide load capacity of 10^4 N when running at 40 rev/s with oil of viscosity grade 46 (ISO 3448):

    From Fig. 16.2, $W_b = 3.8 \times 10^4$ N.
    From Fig. 16.3, viscosity factor 0.75.

    Necessary diameter ratio factor
$$= \frac{10^4}{3.8 \times 10^4 \times 0.75} = 0.35$$

From Fig. 16.4, D/d required is 1.57.
Therefore, outer diameter D required is 157 mm.

Power loss

For a bearing designed in accordance with the recommendations given, the power loss may be estimated by:

$$H = K_1 \, K_2 \, K_3 \, d^3 \, n^{1.66}$$

where the symbols have the meanings given in the following table.

Symbol	Meaning	SI units	Imperial units
d	Inner diameter	mm	in
n	Speed	rev/s	r.p.s.
H	Power loss	W	h.p.
K_1		5.5×10^{-6}	0.12×10^{-3}
K_2	Table 16.2		
K_3	Table 16.3		

Table 16.2 Viscosity grade factor for power loss

Viscosity grade ISO 3448	K_2
32	0.64
46	0.78
68	1.0
100	1.24

Table 16.3 Diameter ratio factor for power loss

D/d	K_3
1.5	0.42
1.6	0.54
1.7	0.66
1.8	0.80
1.9	0.95
2.0	1.1
2.2	1.5
2.4	1.9
2.6	2.4

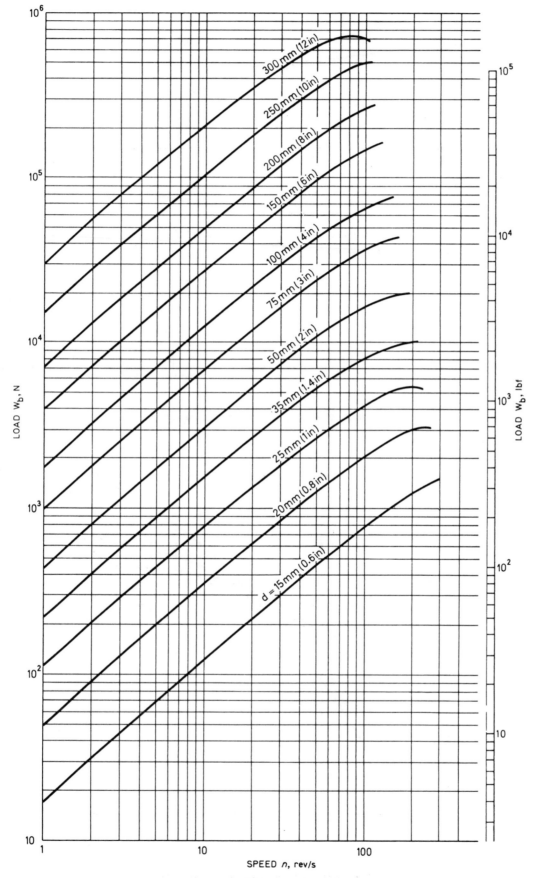

SPEED *n*, rev/s

Fig. 16.2. Basic load capacity W_b

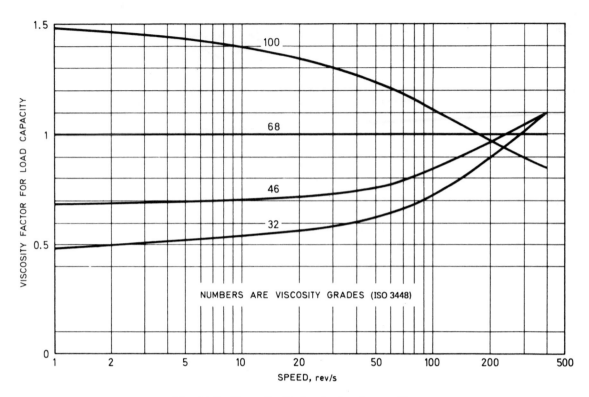

Fig. 16.3. Viscosity factor for load capacity

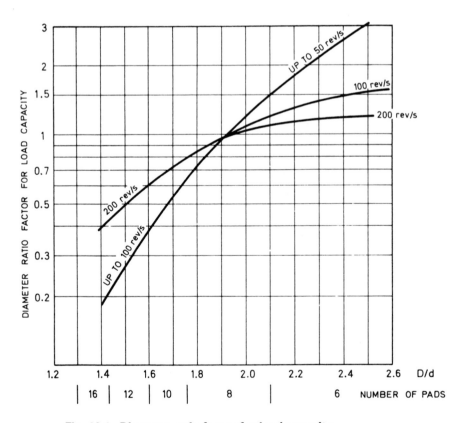

Fig. 16.4. Diameter ratio factor for load capacity

LUBRICANT FEEDING

Lubricant should be directed to the inner diameter of the bearing so that it flows radially outward along the oil-ways. The outlet from the bearing housing should be arranged to prevent oil starvation at the pads.

At high speed, churning power loss can be very significant and can be minimised by sealing at the shaft and runner periphery to reduce the area of rotating parts in contact with lubricant.

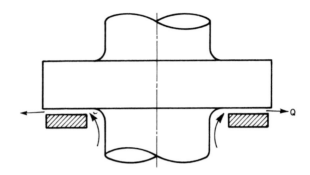

Lubricant feed rate

A lubricant temperature rise of 20°C in passing through the bearing housing is typical. For a feed temperature of 50°C the housing outlet temperature will then be 70°C, which is satisfactory for general use with hydrocarbon lubricants. The flow rate necessary for 20°C temperature rise may be estimated by

$$Q = KH,$$

where $K = 0.3 \times 10^{-7}$ for Q in m³/s, H in W
or $K = 0.3$ for Q in gal/min, H in h.p.

BEARINGS FOR BI-DIRECTIONAL OPERATION

For bi-directional operation a tapered region at both ends of each pad is necessary. In consequence each pad should be circumferentially longer than the corresponding uni-directional pad. The ratio (mean circumferential length/radial width) should be about 1.7 with central land 20% of length. This results in a reduction of the number of pads in the ring, i.e. about $\frac{2}{3}$ the number of pads of the corresponding uni-directional bearing. The resulting load capacity will be about 65%, and the power loss about 80%, of the corresponding uni-directional bearing.

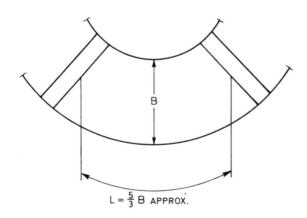

$L = \frac{5}{3} B$ APPROX.

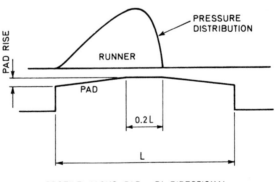

PROFILE ALONG PAD — BI-DIRECTIONAL

The tilting pad bearing is able to accommodate a large range of speed, load and viscosity conditions because the pads are pivotally supported and able to assume a small angle relative to the moving collar surface. This enables a full hydrodynamic fluid film to be maintained between the surfaces of pad and collar. The general proportions and the method of operation of a typical bearing are shown in Fig. 17.1. The pads are shown centrally pivoted, and this type is suitable for rotation in either direction.

Each pad must receive an adequate supply of oil at its entry edge to provide a continuous film and this is usually achieved by immersing the bearing in a flooded chamber. The oil is supplied at a pressure of 0.35 to 1.5 bar ($5-22\,\text{lbf/in}^2$) and the outlet is restricted to control the flow. Sealing rings are fitted at the shaft entry to maintain the chamber full of oil. A plain journal bearing may act also as a seal. The most commonly used arrangements are shown in Fig. 17.2.

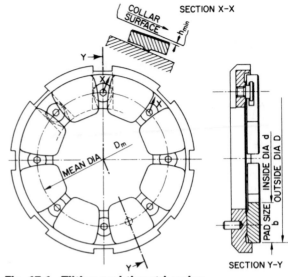

Fig. 17.1. Tilting pad thrust bearing

Fig. 17.2. Typical mounting and lubrication arrangements

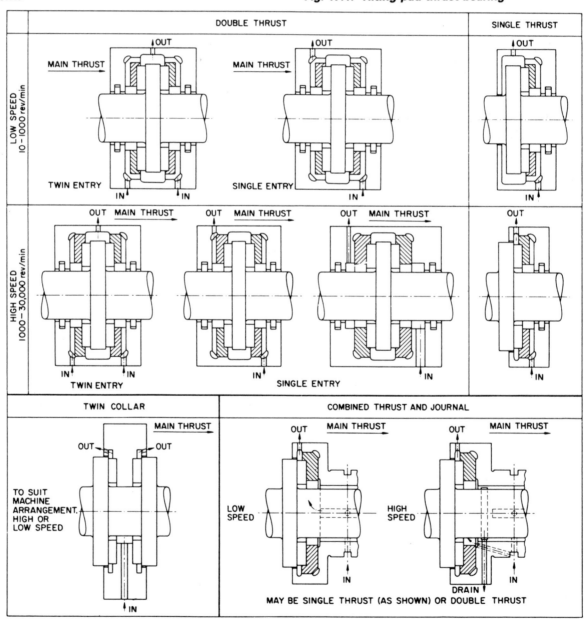

SELECTION OF THRUST BEARING SIZE

The load carrying capacity depends upon the pad size, the number of pads, sliding speed and oil viscosity. Using Figs. 17.3–17.7 a bearing may be selected and its load capacity checked. If this capacity is inadequate then a reiterative process will lead to a suitable bearing.

(1) Use Fig. 17.3(a) for first approximate selection.
(2) From Fig. 17.3(b) select diameters D and d.
(3) From Fig 17.4 find thrust ring mean diameter D_m.
(4) From Fig. 17.5 find sliding speed.
(5) For the bearing selected calculate:

$$\text{Specific load, } p \text{ (MN/m}^2) = \frac{\text{Thrust load (N)}}{\text{Thrust surface (mm}^2)}$$

Check that this specific load is below the limits set by Figs. 17.6(a) and 17.6(b) for safe operation.

Note: these curves are based on an average turbine oil having a viscosity of 25 cSt at 60°C, with an inlet to the bearing at 50°C.

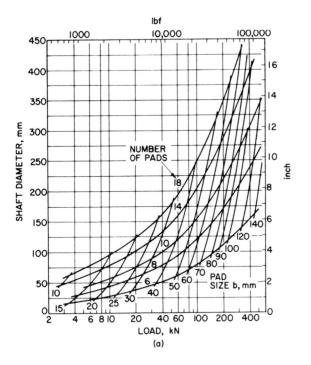

(a)

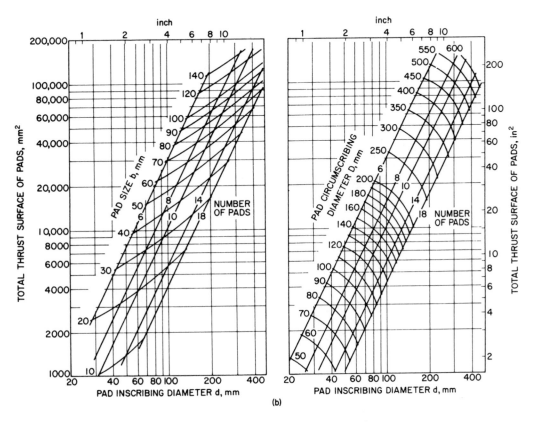

(b)

Fig. 17.3. First selection of thrust bearing

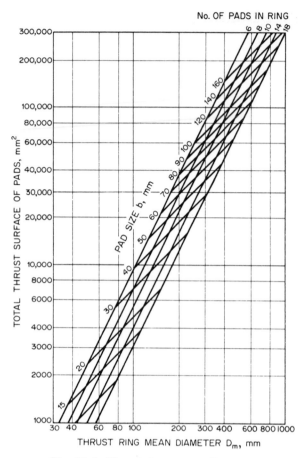

Fig. 17.4. Thrust ring mean diameter

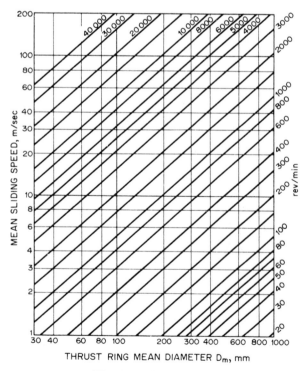

Fig. 17.5. Sliding speed

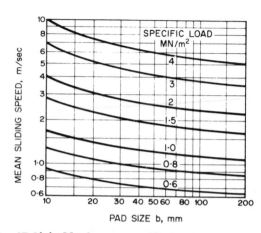

Fig. 17.6(a). Maximum specific load at slow speed to allow an adequate oil film thickness

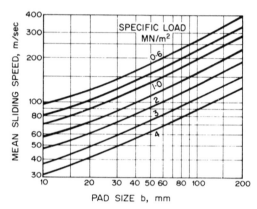

Fig. 17.6(b). Maximum specific load at high speed to avoid overheating in oil film

The load carrying capacity varies with viscosity, and for different oils may be found by applying the correction factors in Fig. 17.7:

Maximum specific load = Specific load (Fig. 17.6)
$\times$ factor f (Fig. 17.7).

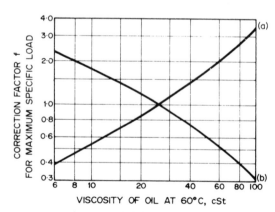

Fig. 17.7. Maximum safe specific load:
slow speed = f (curve (a)) $\times$ spec. load Fig. 17.6(a)
high speed = f (curve (b)) $\times$ spec. load Fig. 17.6(b)

CALCULATION OF POWER ABSORBED

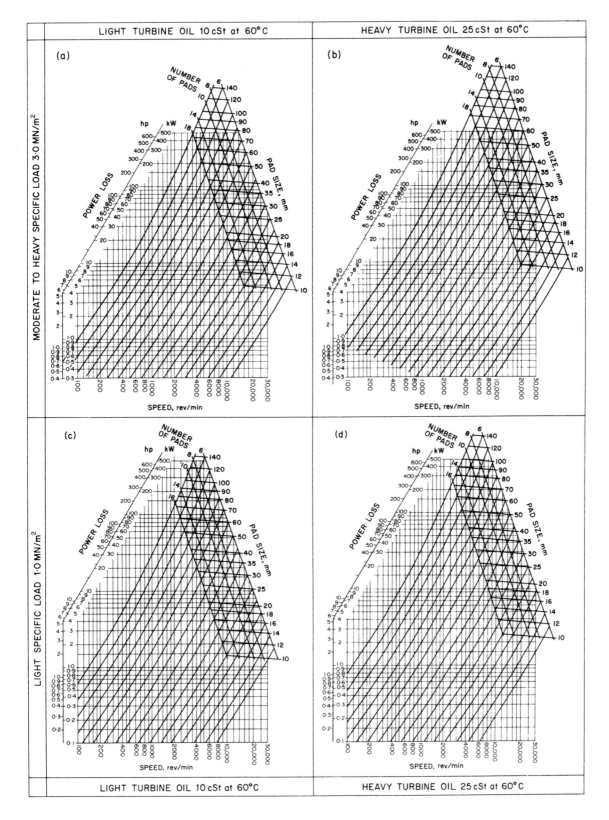

Fig. 17.8. Power loss guidance charts, double thrust bearings

The total power absorbed in a thrust bearing has two components:

(1) Resistance to viscous shear in the oil film.
(2) Fluid drag on exposed moving surfaces—often referred to as 'churning losses'.

The calculations are too complex to be included here and data should be sought from manufacturers. Figure 17.8 shows power loss for typical double thrust bearings.

Figure 17.9 shows the components of power loss and their variation with speed.

Note: always check that operating loads are in the safe region.

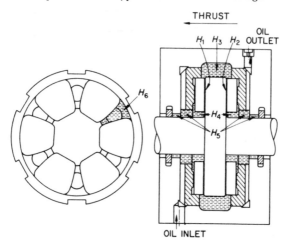

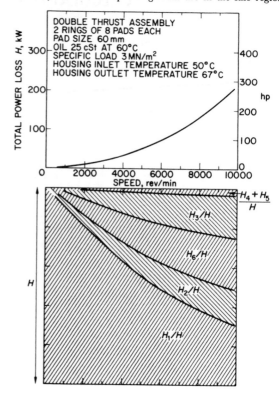

H = total power loss
H_1 = film shear at main face
H_2 = film shear at surge face
H_3 = drag loss at rim of collar
H_4 = drag loss at inside of pads
H_5 = drag loss along shaft
H_6 = drag loss between pads

Fig. 17.9 Components of power loss in a typical double thrust bearing

High speed bearings

In low speed bearings component 2, the churning loss, is a negligible proportion of the total power loss but at high speeds it becomes the major component. It can be reduced by adopting the arrangement shown in Fig. 17.10. Instead of the bearing being flooded with oil, the oil is injected directly on to the collar face to form the film. Ample drain capacity must be provided to allow the oil to escape freely.

Figure 17.11 compares test results for a bearing having 8 pads, $b = 28$ mm, at an oil flow of 45 litres/min.

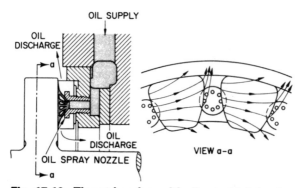

Fig. 17.10. Thrust bearing with directed lubrication

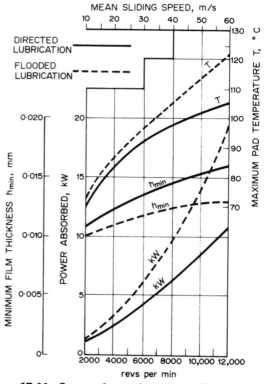

Fig. 17.11. Comparison between flooded and directed lubrication

Oil flow

Oil is circulated through the bearing to provide lubrication and to remove the heat resulting from the power loss.

It is usual to supply oil at about 50°C and to allow for a temperature rise through the bearing of about 17°C.

There is some latitude in the choice of oil flow and temperature rise, but large deviations from these figures will affect the performance of the bearing.

The required oil flow may be calculated from the power loss as follows:-

$$\text{Oil flow (litres/min)} = 35.8 \times \frac{\text{Power Loss (kW)}}{\text{Temperature Rise (°C)}}$$

$$\text{Oil Flow (US gals/min)} = 12.7 \times \frac{\text{Power Loss (hp)}}{\text{Temperature Rise (°F)}}$$

EQUALISED PAD BEARINGS

Where the bearing may be subject to misalignment, either due to initial assembly, or to deflection of the supporting structure under load an alternative construction can be adopted, although with the disadvantages of increased size and expense.

The equalised pad bearing is shown in Fig. 17.12. The pads are supported on a system of interlinked levers so that each pad carries an equal share of the load.

Misalignment of the order of up to 0.1° (0.002 slope) can be accepted. Above this the equalising effect will diminish.

In practice the ability to equalise is restricted by the friction between the levers, which tend to lock when under load. Thus the bearing is better able to accept initial misalignment than deflection changes under load.

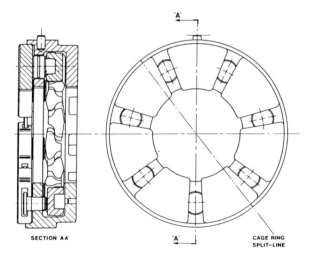

SECTION 'AA'   'A'   CAGE RING SPLIT-LINE

Fig. 17.12. An equalised pad thrust bearing

STARTING UNDER LOAD

In certain applications, notably vertical axis machines the bearing must start up under load. The coefficient of friction at break-away is about 0.15 and starting torque can be calculated on the Mean Diameter.

The specific load at start should not exceed 70% of the maximum allowable where acceleration is rapid and 50% where starting is slow.

Where load or torque at start are higher than acceptable, or for large machines where starting may be quite slow a jacking oil system can be fitted. This eliminates friction and wear.

BEARINGS FOR VERY HIGH SPEEDS AND LOADS

Traditionally the thrust pads are faced with whitemetal and this is still the most commonly used material. But, with increasingly higher specific loads and speeds the pad surface temperature will exceed the permissible limit for whitemetal – usually a design temperature of 130°C.

Two alternative approaches are available:-

1   The pad temperature may be reduced by

   (a)   Directed lubrication – see Figs. 17.10 and 17.11.

   (b)   Adopting offset pivots; accepting their disadvantages.

   (c)   Changing the material of the pad body to high conductivity. Copper – Chromium alloy.

2   Alternatively the pads can be faced with materials able to withstand higher temperatures but at increased cost.

   (a)   **40% Tin–Aluminium** will operate 25°C higher than whitemetal. Has comparable boundary lubrication tolerance and embeddability with better corrosion resistance.

   (b)   **Copper–Lead** will operate 40°C higher than whitemetal. Poorer tolerance to boundary lubrication and embeddability. Requires the collar face to be hardened.

   (c)   **Polymer based upon PEEK** can be used at temperatures up to 200°C and above. Comparable embeddability and better tolerance to boundary lubrication. Suitable for lubrication by water and mainly low viscosity process fluids.

   (d)   **Ceramic Pads and Collar Face**, made from silicon – carbide, these can be used up to 380°C and specific loads up to about 8 MPa (1200 p.s.i.). They are chemically inert and suitable for lubrication by low viscosity fluids such as water, most process fluids and liquified gases.

LOAD MEASUREMENT

The bearings can be adapted to measure thrust loads using either electronic or hydraulic load cells. The latter can provide very effective load equalisation under misalignment and may be used to change the axial stiffness at will to avoid resonant vibration in the system.

Hydrostatic bearings

In a hydrostatic bearing the surfaces are separated by a film of lubricant supplied under pressure to one or more recesses in the bearing surface. If the two bearing surfaces are made to approach each other under the influence of an applied load the flow is forced through a smaller gap. This causes an increase in the recess pressure. The sum of the recess pressure and the pressures across the lands surrounding the recess build up to balance the applied load. The ability of a bearing film to resist variations in gap with load depends on the type of flow controller.

LOAD CAPACITY

Figures 18.1 and 18.2 give an approximate guide to the load capacities of single plane pads and journal bearings at various lubricant supply pressures. Approximate rules are:
(1) The maximum mean pressure of a plane pad equals one-third the supply pressure.
(2) The maximum mean pressure on the projected area of journal bearings and opposed pad bearings equals one-quarter the supply pressure.

An approximate guide to average stiffness λ may be obtained by dividing the approximate load by the design film thickness $\lambda = W/h_o$.

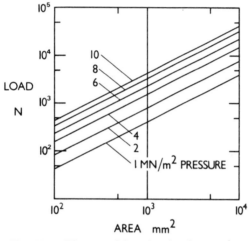

Fig. 18.1. Plane pad bearing load capacity

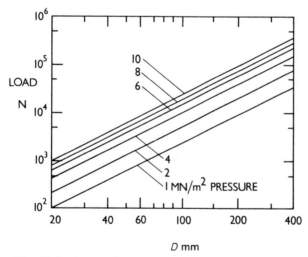

Fig. 18.2. Journal bearing load capacity

CONTROL CIRCUITS

Each recess in a bearing must have its own flow controller, as illustrated in Figs. 18.3 and 18.4, so that each recess may carry a load independently of the others.

Flow controllers in order of increasing bearing stiffness are as follows:
(1) Laminar restrictors (capillary tubing).
(2) Orifices (length to diameter ratio $\ll 1$).
(3) Constant flow (fixed displacement pumps or constant flow valves).
(4) Pressure sensing valves.

Figure 18.4(a) illustrates a typical circuit for capillary or orifice control. The elements include a filter (FLT), a motor (M), a fixed displacement pump (PF), an inlet strainer (STR), and a flow relief valve set to maintain the operating supply pressure at a fixed maximum p_f. Figure 18.4(b) shows a circuit involving a constant flow control valve with pressure compensation (PC).

The control circuit must be designed to provide the necessary value of recess pressure p_o at the design bearing clearance h_o. It is first necessary to calculate the flow from the bearing recess at the design condition Q_o. The values of Q_o and p_o are then employed in the calculation of the restrictor dimensions and in selection of other elements in the circuit.

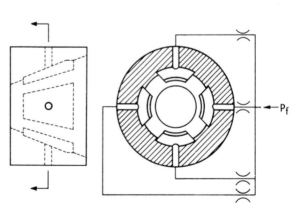

Fig. 18.3. A conical hydrostatic journal bearing showing recesses and restrictions

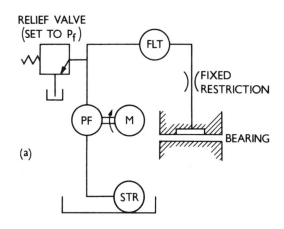

(a)

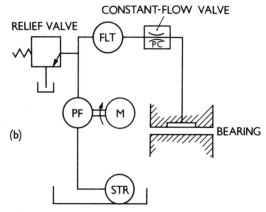

(b)

Fig. 18.4. Layout of typical control circuits

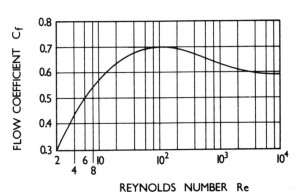

Fig. 18.5. Values of K_c for capillaries of various dimensions

Design of control restrictors

Dimensions of capillary and orifice restrictors may be calculated from the following equations

Capillary flow $\qquad Q_o = \dfrac{p_f - p_o}{K_c \eta}$

where $\qquad\qquad K_c = \dfrac{128\, l}{\pi d^4}$

Suitable values for length and diameter are presented in Fig. 18.5. The l/d ratio should preferably be greater than 100 for accuracy and the Reynolds Number should be less than

1000 where $R_e = \dfrac{\rho v d}{\eta}$, ρ = density, v = average velocity,

d = bore diameter and η = dynamic viscosity.

Orifice flow $Q_o = \dfrac{\rho}{2(C_f A)^2}\left(p_f - p_o\right)^{\frac{1}{2}}$

$\qquad A$ = cross-sectional area of the orifice.

Values of the flow coefficient C_f are presented in Fig. 18.6 according to the values of R_e.

Fig. 18.6. Values of the flow coefficient C_f for orifices

87

Calculation of bearing stiffness

Bearing characteristics are dependent not only on the type of control device. The characteristics are also dependent on the design pressure ratio $\beta = \dfrac{p_o}{p_f}$. Figures 18.8 and 18.9 show the variation of dimensionless stiffness parameter $\bar{\lambda}$ with the film thickness h and design pressure ratio β. Taking all considerations into account including manufacturing tolerances it is recommended to aim for $\beta = 0.5$. The relationship between stiffness λ and the dimensionless value $\bar{\lambda}$ is

$$\lambda = \frac{P_f A_e}{h_o} . \bar{\lambda}$$

where A_e is the effective area of the pad over which p_f may be assumed to act. For a plane pad in which the recess occupies one-quarter of the total bearing pad area A the effective area is approximately $A/2$ which may be deduced by assuming that p_f extends out to the mid-land boundary. For more accuracy the effective area may be expressed as $A_e = A\bar{A}$ which defines a dimensionless area factor $\bar{A}$. Some computed values of $\bar{A}$ are presented for plane pads in Figs. 18.7 and 18.10.

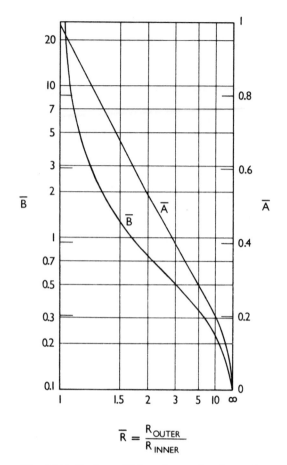

$$\bar{R} = \frac{R_{OUTER}}{R_{INNER}}$$

Fig. 18.7. Pad coefficient for a circular pad

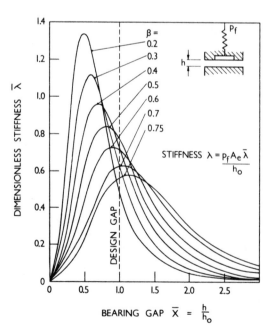

Fig. 18.8. Stiffness parameters for capillary-compensated single pad bearings

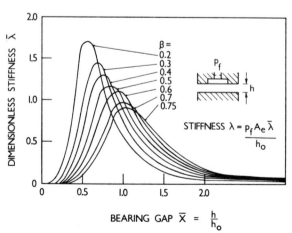

Fig. 18.9. Stiffness parameters for orifice-controlled single pad bearings

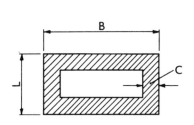

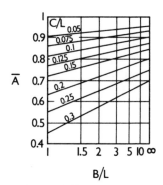

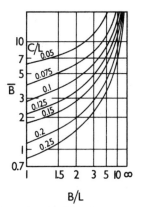

Fig. 18.10. Pad coefficient for a rectangular pad. For a rectangular pad with uniform land width it is recommended that C / L < 0.25

PLANE HYDROSTATIC PAD DESIGN

The performance of plane pad bearings may be calculated from the following formulae:

Load: $$W = p_f A . \bar{A} . \bar{P}$$

Flow: $$Q = \frac{p_f h_o^3}{\eta} . \bar{P} . \bar{B}$$

where $\bar{A}$ is a factor for effective area $(A_e = A\bar{A})$
$\bar{B}$ is a factor for flow

$\bar{P} = \dfrac{p}{p_f}$ and varies with film thickness

$h_o = $ design film thickness

Figures 18.7 and 18.10 give values of $\bar{A}$ and $\bar{B}$ for circular and rectangular pads of varying land widths.

The relationship between $\bar{P}$ and h depends on β and curves are presented in Figs. 18.11 and 18.12 for capillary and orifice control.

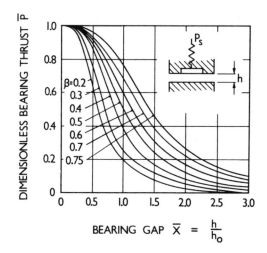

Fig. 18.11. Variation of thrust pad load capacity with film thickness using capillary control

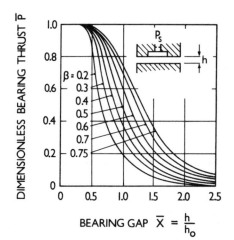

Fig. 18.12. Variation of thrust pad load capacity with film thickness using orifice control

For bearings which operate at speed it is important to optimise the design to minimise power dissipation and to prevent cavitation and instability problems. The optimisation required may be achieved by selecting values of viscosity η and film thickness h_o to satisfy the following equation:

$$\frac{\eta v}{p_f h_o^2} = \left(\frac{\beta \bar{B}}{A_f}\right)^{\frac{1}{4}}$$

where $A_f = $ (total area) $-\frac{3}{4}$ (recess area)

$\qquad\quad = $ effective friction area

Recess depth $= 20 \times$ bearing clearance h_o

$v = $ linear velocity of bearing

The above relationship minimises total power which is the sum of friction power and pumping power. A further advantage of optimisation is that it ensures that temperature rise does not become excessive. For optimised bearings the maximum temperature rise as the lubricant passes through the bearing may be calculated from

$$\Delta T = \frac{2 p_f}{\mathcal{J} \rho C_v}$$

where $\mathcal{J} = $ mechanical equivalent of heat

$\qquad C_v = $ specific heat

DESIGN OF HYDROSTATIC JOURNAL BEARINGS

The geometry and nomenclature of a cylindrical journal bearing with n pads are illustrated in Fig. 18.13. For journal bearings the optimum value of design pressure ratio is $\beta = 0.5$ as for other hydrostatic bearings. Other values of β will reduce the minimum film thickness and may reduce the maximum load. The following equations form a basis for safe design of journal bearings with any number of recesses and the three principal forms of flow control (refer to Fig. 18.13 and Table 18.1).

Load: $W = p_f A_e . \bar{W}$

where $\bar{W}$ is a load factor which normally varies from

0.30 to 0.6 a better guide is $\bar{W} = \dfrac{\bar{\lambda}'}{2}$

$\bar{\lambda}$ = dimensionless stiffness parameter from Table 18.1
$\bar{\lambda}'$ = value of $\bar{\lambda}$ for capillary control and $\beta = 0.5$,
$A_e = D(L-a)$.

Concentric stiffness: $\quad \lambda = \dfrac{p_f A_e}{C} . \bar{\lambda}$

where $\qquad C = h_o$ = radial clearance.

Flow-rate: $\qquad Q = \dfrac{p_f C^3}{\eta} . n \beta \bar{B}$

where $\qquad \bar{B} = \dfrac{\pi D}{6 a n}$

is the flow factor for one of the n recesses.

$$\gamma = \frac{n a (L-a)}{\pi D b}$$

is a circumferential flow factor. If the dimension 'b' is too small the value γ will be large and the bearing will be unstable.

The recommended geometry for a journal bearing (see Fig. 18.13).

$$a = \frac{L}{4}, \quad \frac{L}{D} = 1, \quad b = \frac{\pi D}{4 n}$$

Journal bearings which operate at speed should be optimised for minimum power dissipation and low temperature rise for the same reasons as given under the

Table 18.1 Dimensionless stiffness $\bar{\lambda}$ (for a journal bearing with n pads)

n	Capillary	Orifice	Constant flow
4	$\dfrac{3.82\,\beta\,(1-\beta)}{1+\gamma\,(1-\beta)}$	$\dfrac{7.65\,\beta\,(1-\beta)}{2-\beta+2\,\gamma\,(1-\beta)}$	$\dfrac{3.82\,\beta}{1+\gamma}$
5	$\dfrac{4.12\,\beta\,(1-\beta)}{1+0.69\,\gamma\,(1-\beta)}$	$\dfrac{8.25\,\beta\,(1-\beta)}{2-\beta+1.38\,\gamma\,(1-\beta)}$	$\dfrac{4.25\,\beta}{1+0.69\,\gamma}$
6	$\dfrac{4.30\,\beta\,(1-\beta)}{1+0.5\,\gamma\,(1-\beta)}$	$\dfrac{8.60\,\beta\,(1-\beta)}{2-\beta+\gamma\,(1-\beta)}$	$\dfrac{4.30\,\beta}{1+0.5\,\gamma}$

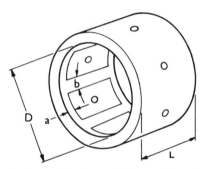

Fig. 18.13. Typical hydrostatic journal bearing

previous paragraph headed 'Plane Hydrostatic Pad Design'. Values of viscosity and clearance should be selected so that:

$$\frac{\eta \, \mathcal{N}'}{p_f} \left(\frac{D}{C_D}\right)^2 = \frac{1}{4\,\pi} \left(\frac{n\,\beta\,\bar{B}}{\bar{A}_f}\right)^{\frac{1}{2}}$$

where $\mathcal{N}'$ = rotational speed in rev/sec

$\bar{A}_f = [(\text{total area}) - \frac{3}{4}(\text{recess area})]/D^2$

Recess depth = $20 \times$ radial clearance. Maximum temperature rise may be calculated as for plane pads.

EXTERNALLY PRESSURISED GAS BEARINGS

Externally pressurised gas bearings have the same principle of operation as hydrostatic liquid lubricated bearings. There are three forms of external flow restrictor in use.

Restrictor type		Pocketed orifice (simple orifice)	Unpocketed orifice (annular orifice)	Slot
Flow restriction		$\alpha \dfrac{1}{\pi d^2/4}$	$\alpha \dfrac{1}{\pi dh}$	$\alpha \dfrac{-y}{az^3}$
Geometrical restrictions		$\dfrac{d^2}{4} < d\delta$ i.e. $\delta > \dfrac{d}{4}$ $\dfrac{\pi d^2}{4} < \pi d_p h$	When $\dfrac{\pi d^2}{4} < \pi dh$ i.e. $d < 4h$ orifice reverts to simple type giving increased bearing stiffness	y is generally large for practical values of z resulting in large overall bearing size
Steady state load capacity	Journal and back-to-back thrust	Generally the highest	33% less than pocketed	Highest for very low b/D journals
	Single thrust	Can approach that of slot type by spreading pressure with surface grooves	Same as pocketed	Generally the highest
Steady state stiffness K	Journal and back-to-back thrust	Generally the highest	33% less than pocketed	Highest for very low b/D journals
	Single thrust	Highest	33% less than pocketed	
Damping		Both pockets and surface grooves will reduce damping		
Stability		Both pockets and surface grooves can result in instabilities		
Dynamics		Overall damping decreases with rotation in journal bearings. All bearings are subject to resonance at the natural frequency $\omega_n = \sqrt{(K/m)}$: symmetrical rigidly supported journal systems have 'whirl' at a speed $2 \times \omega_n$		
Lubricant		Any clean gas: design charts presented are for air.		

Externally pressurised gas thrust bearings

There are two main types of externally pressurised gas thrust bearings. These are the central recess and annular bearing as shown on the right. The central recess bearing is fed from a central orifice, while the annular bearing is fed from a ring of orifices. The stiffness of these bearings varies with clearance as shown below.

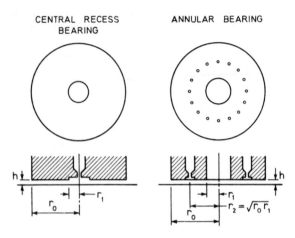

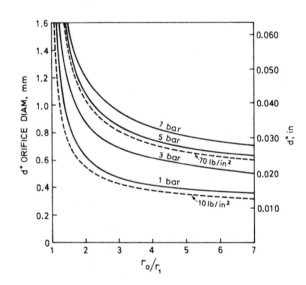

Orifice dia. $d*$ for a central recess bearing with clearance $h_0 = 25 \ \mu m$ (0.001 in) for *air*, 15°C, ambient pressure 1 bar, is given below.

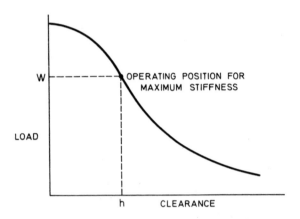

GENERAL LOAD CLEARANCE CHARACTERISTIC

All design data are given for operation at the position of maximum stiffness and are approximate.

Load carried $W = C_L' \cdot \pi r_0^2 \, P_f$

where $P_f = $ supply pressure (gauge)

Stiffness $K = 1.42 \ W/h$ for pocketed orifices

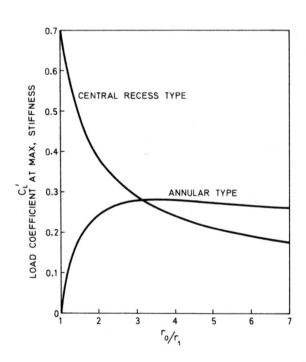

For other values of clearance h new orifice dia.;

$$d = d* \left(\frac{h}{h_0} \right)^{\frac{3}{4}}$$

For annular bearing with N pocketed orifices;

$$d = d* \cdot \left(\frac{h}{h_0} \right)^{\frac{3}{4}} \cdot \frac{2}{\sqrt{N}}$$

For annular bearing with N unpocketed orifices;

$$d = \frac{d*^2 \cdot h^2}{N h_0^3}$$

and stiffness $K = 0.95 \dfrac{W}{h}$

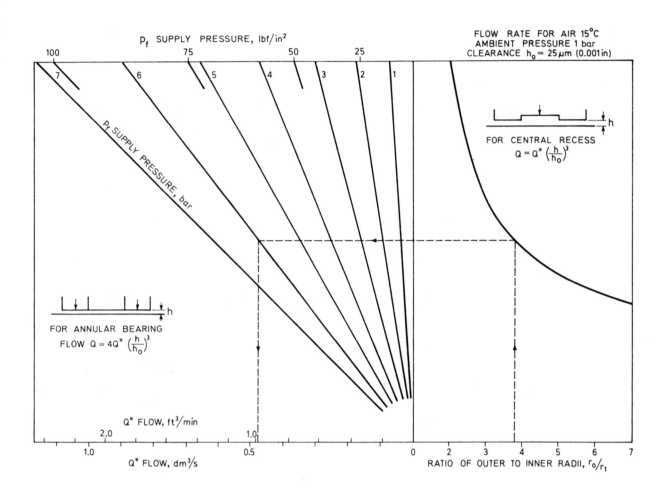

Example of thrust bearing design

Load to be carried weighs 1000 N. Bearing must have central hole 50 mm dia.: 5 bar supply pressure available. Design for maximum stiffness.

Take maximum load coefficient available as starting point. $C_L' = 0.28$. Then since $W = 1000$ N; $P_f = 5$ bar;

$$r_0^2 = \frac{1000}{0.28\pi \cdot 5.10^5} = 2.10^{-3} \text{ m}^2,$$

i.e. $r_0 = 0.045$ m and $r_0/r_1 = 0.045/0.025 = 1.8$. But C_L' was taken at $r_0/r_1 = 3$: bearing cannot operate at maximum C_L' with 5 bar supply pressure. Design can therefore be off maximum C_L' or at lower supply pressure.

(1) Take $r_0/r_1 = 2$, then $C_L' = 0.25$, $r_0 = 50$ mm: $W = 0.25 \quad \pi \quad (0.050)^2, \quad 5.10^5 = 1000$ N. To operate with 20 μm clearance; then from d^* graph, $d^* = 1.08$ mm. Thus for this annular bearing with say, 16 orifices; orifice diameter

$$d = 1.08 \left(\frac{20}{25}\right)^{\frac{3}{2}} \times \frac{2}{(16)^{\frac{1}{2}}} = 0.4 \text{ mm}:$$

$$\text{Stiffness} = \frac{1.42 \times 1000}{0.020} = 70 \text{ kN/mm}$$

or; for greater stability operate with unpocketed orifices. Then for $N = 40$ say,

$$d = \frac{1.08^2}{40} \frac{0.020^2}{0.025^3} = 0.75 \text{ mm [check } d > 4h]:$$

$$\text{Stiffness} = \frac{0.95 \times 1000}{0.020} = 46 \text{ kN/mm.}$$

Flow rate is not dependent upon orifice type, thus flow rate for either type

$$Q = 4Q* \left(\frac{20}{25}\right)^2,$$

now $Q* = 0.65$ dm^3/s, therefore

$$Q = 4 \times 0.65 \times 0.51 = 1.3 \text{ dm}^3/\text{s.}$$

(2) For $r_0/r_1 = 3$, i.e. $r_0 = 75$ mm, and $C_L' = 0.28$,

$$P_f = \frac{1000}{0.28 \cdot \pi (0.075)^2} = 2 \text{ bar;}$$

for operation at 20 μm clearance stiffness is unchanged from (1). Flow $Q* = 0.12$ dm^3/s, therefore actual flow rate

$$Q = 4 \times 0.12 \left(\frac{20}{25}\right)^3 = 0.25 \text{ dm}^3/\text{s,}$$

cf. 1.3 dm^3/s for arrangement (1).

Externally pressurised—gas journal bearings

This section gives an approximate guide to load carrying capacity and flow requirements for a design which is optimised for load capacity and stiffness.

Values are given for a bearing with two rows of 8 orifices as shown on the right, and with $l/b = \frac{1}{4}$, $P_f^* = 6.9$ bar (100 p.s.i.).

The load W^* at an eccentricity $\varepsilon = \dfrac{2e}{C_d} = 0.5$ is given below.

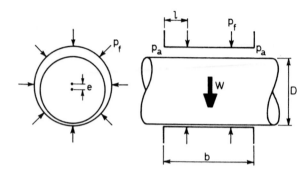

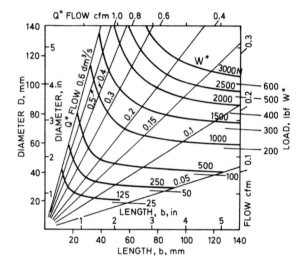

Load for other supply pressures for pocketed orifices

$$W = W^* \frac{P_f}{P_f^*}$$

Load at $\varepsilon = 0.5$ for unpocketed orifices

$$W = \tfrac{2}{3} W^* \frac{P_f}{P_f^*}$$

for higher load capacity:
(load at $\varepsilon = 0.9$) = 1.28 (load at $\varepsilon = 0.5$).

$$\text{Stiffness} = \frac{(\text{load at } \varepsilon = 0.5)}{C_d/4}$$

Flow Q^* is given for $C_d^* = 25\ \mu\text{m}$ (0.001 in), air, 15°C, $p_a = 1$ bar.

At other clearances, flow $Q = Q^* \left(\dfrac{C_d}{C_d^*}\right)^3$.

For other numbers of orifices per row use graph below.

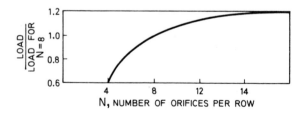

Orifice size d^* is given for:

$b/D = 1$, 8 orifices per row
air, 15°C, $p_a = 1$ bar
$C_d^* = 25\ \mu\text{m}$ (0.001 in)

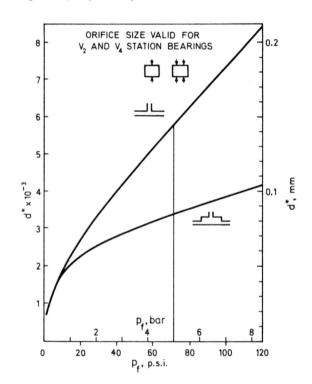

Under other operating conditions,
for pocketed orifices:

$$d = d^* \cdot \left(\frac{C_d}{C_d^*}\right)^{\frac{1}{4}} \cdot \left(\frac{8}{N}\frac{D}{b}\right)^{\frac{1}{4}}$$

for unpocketed orifices

$$d = d^* \cdot \frac{C_d}{C_d^*} \cdot \frac{8D}{Nb}$$

In general, orifices are positioned at either $l/b = \frac{1}{4}$ (quarter station configuration).
or a single row at $l/b = \frac{1}{2}$ (half station)

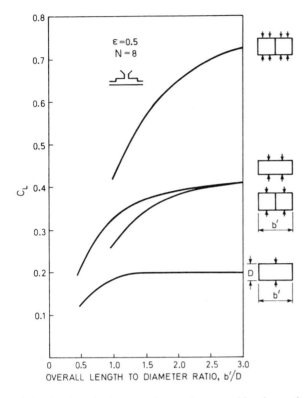

OVERALL LENGTH TO DIAMETER RATIO, b'/D

The load coefficient C_L for various combinations of half and quarter stations is used to assess the best arrangement for the highest load capacity bearing for a fixed shaft diameter. Load carrying capacity at $\varepsilon = 0.5$ and 8 orifices per row (pocketed) $W = C_L . P_f . D^2$.
For low pumping power use half stations. For greater load capacity and stiffness from smallest area bD use quarter stations.

Example of journal bearing design

Design wanted for spindle bearing. Radial stiffness must be greater than 200 kN/mm to achieve required resonant frequency; 5 bar supply pressure is available. Minimum shaft dia. set by shaft stiffness is 100 mm and minimum practicable clearance $C_d = 30$ μm. What is the shortest length of bearing which can be used?

(i)
$$\text{Stiffness} = \frac{W}{C_d/4},$$

therefore bearing can be defined by load which can be carried at $\varepsilon = 0.5$,

i.e. $\qquad W = (200.10^6) \dfrac{(30.10^{-6})}{4} = 1500$ N.

Using minimum shaft dia., $D = 100$ mm and if $P_f = 5$ bar
$$C_L = \frac{W}{P_f D^2} = \frac{1500}{(5.10^5)(0.1)^2} = 0.3$$

Therefore the shortest bearing system will be, a single $\frac{1}{4}$ station bearing with $b/D = 0.8$, i.e. $b = 80$ mm from the graph of C_L. Orifice size for this bearing is obtained via the graph of $d*$, $d* = 0.085$ mm for $b/D = 1$ and $C_d^* = 25$ μm and 8 orifices per row.
With our values for b/D and C_d;

$$d = 0.085 \left(\frac{30}{25}\right)^{\frac{3}{2}} \left(\frac{1}{0.8}\right)^{\frac{1}{4}} = 0.125 \text{ mm.}$$

Use of two short bearings side by side can give higher load capacity and stiffness for given length but with greatly increased air flow.
Air flow rate is given for bearing with $b/D = 1$, $l/b = \frac{1}{4}$ and $C_d^* = 25$ μm (0.001 in).

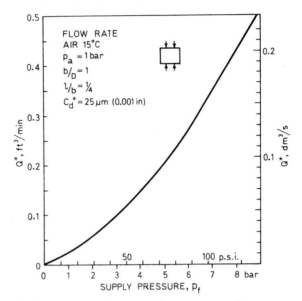

Flow rate for other operating conditions is given by
$$Q = Q* \cdot \left(\frac{C_d}{C_d^*}\right)^3 \cdot \frac{D}{b}$$
For $\frac{1}{2}$ station bearings
$$Q = Q* \left(\frac{C_d}{C_d^*}\right)^3 \frac{D}{2b}$$

Flow rate is not dependent upon orifice type.

This is an acceptable value. We could increase stiffness by increasing the number above 8 per row but may introduce manufacturing problems.

(ii) If unpocketed orifices were required for improved stability, then:

Stiffness $= \dfrac{2}{3} \dfrac{W}{C_d/4}$ $\quad \therefore W = \frac{3}{2} . 1500 = 2250$ N
and $\qquad\qquad\qquad C_L = \frac{3}{2} 0.3 = 0.45$.

We must use two $\frac{1}{4}$ station bearings side by side; overall $b'/D = 1.2$, $b' = 120$ mm. Each bearing has $b/D = 0.6$. $d* = 0.145$. Therefore actual orifice dia. required

$$d = 0.145 \; \frac{30}{25} \; \frac{1}{0.6} = 0.29 \text{ mm}$$

at 8 orifices/row.
Alternatively the shaft dia. could be increased to allow the use of a single $\frac{1}{4}$ station bearing, saving on air flow rate; since a short bearing is required take $C_L = 0.32$ at $b/D = 1$. Then $W = 0.32 \times 5 \times 10^5 \; D^2$ and since $W = 2250$ N,

$$D^2 = \frac{2250}{0.32 \times 5 \times 10^5} = 1.41 \; 10^{-2} \text{ m}^2,$$

$D = 119$ mm, $b = 119$ mm.

Points to note in designing externally pressurised gas bearings

Designing for	Points to note
High speed	Power absorbed and temperature rise in bearing due to viscous shear. Resonance and whirl in cylindrical and conical modes. Variation in clearance due to mechanical and thermal changes
Low friction	Friction torque $= \dfrac{\pi^2 \eta D^3 b n}{C_d}$ Note that flow rate αC_d^3
No inherent torque	Feedholes must be radial. Journal circular and thrust plates without swash. Clearance and supply pressure low to reduce gas velocity in clearance
High load capacity	Spread available pressure over whole of available surface
High stiffness	Need small clearance and pocketed feedholes. In journals and back-to-back thrust bearings use large areas. In single thrust bearings maximum stiffness is proportional to $\dfrac{\text{load}}{\text{clearance}}$ at maximum stiffness condition. The performance of back-to-back thrust bearing arrangements can be deduced from the data given for single sided thrust bearings
Low compressor power	Low clearance, large length/diameter ratio journals and long flow path thrust bearings
High accuracy of rotation	Stator circularity is less important than that of rotor. About 10:1 ratio between mzc value of rotor surface and locus of rotor

SELF-ACTING GAS BEARINGS

1. The mechanism of operation is the same as that of liquid lubricated hydrodynamic bearings.
2. Self-acting gas bearings can be made in a considerable variety of geometric forms which are designed to suit the application.

   The simplest type of journal bearing is a cylindrical shaft running in a mating bearing sleeve. Design data for this type is well documented and reliable. Tilting pad journal and thrust bearings form another well-known group, having higher speed capabilities than the plain journal bearing.

   Another type of self-acting gas bearing uses one form or another of spiral grooves machined into the bearing surfaces. The function of the grooves is to raise the pressure within the bearing clearance due to the relative motion of one surface to the other. Bearings of this type have been applied to cylindrical journals using complementary flat thrust faces. Other shapes that have been frequently used are spherical and conical in form; in which case the bearing surfaces carry both radial and axial load.
3. Gases lack the cooling and boundary lubrication capabilities of most liquid lubricants, so the operation of self-acting gas bearings is restricted by start/stop friction and wear.

   If start/stop is performed under load then the design is limited to about 48 kN/m^2 (7 lbf/in^2) of projected bearing area depending upon the choice of materials. In general the materials used are those of dry rubbing bearings; either hard/hard combination such as ceramics with or without a molecular layer of boundary lubricant or a hard/soft combination using a plastics surface.
4. It is vital that the steady state and dynamic loads being applied to a self-acting gas bearing are known precisely since if contact between the bearing surfaces results from the response to external forces or whirl instabilities then failure of the bearing may occur. Any design should therefore include a dynamic analysis of the system of bearings and rotor.
5. For full design guidance it is advisable to consult specialists on gas bearings or tribology.

DETERMINATION OF BASIC DYNAMIC LOADING RATING

The following nomogram can be used for determining the ratio

$$\frac{C}{P} = \frac{\text{basic dynamic load rating of the required bearing}}{\text{equivalent load to be carried by the bearing}}$$

from the rotational speed and the required rating life. The International Organisation for Standardisation (ISO) defines the rating life of a group of apparently identical rolling bearings as that completed or exceeded by 90% of that group before first evidence of fatigue develops. The median life is approximately 5 times the rating life.

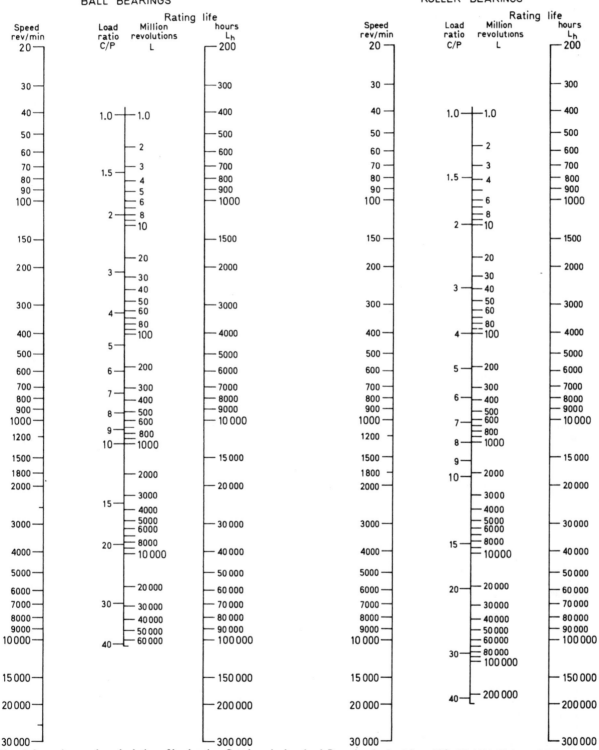

BALL BEARINGS

ROLLER BEARINGS

Note: information on the calculation of load rating C and equivalent load P can be obtained from ISO 281:1990. Values of C for various types of bearing can be obtained from the bearing manufacturers.

SELECTION OF TYPE OF BEARING REQUIRED

Table 20.1 Guide for general application

Design of bearing	*Bearing bore*	*Sealed* (Se) *or shielded* (S)	*Load capability*		*Allowable misalignment (degrees)*[4]	*Coefficient of friction*[5]	*Bearing section*
			Radial	*Axial*			
Single row deep groove ball	Cylindrical	1 or 2 Se 1 or 2 S	Light and medium	Light and medium	0.01 to 0.05	0.0015	
Self-aligning double row ball	Cylindrical or tapered	2 S	Light and medium	Light	2 to 3	0.0010	
Angular contact single row ball	Cylindrical	—	Medium[1]	Medium and heavy	—	0.0020	
Angular contact double row ball	Cylindrical	—	Medium	Medium	—	0.0024	
Duplex	Cylindrical	—	Light[2]	Medium	—	0.0022	
Cylindrical roller single row	Cylindrical	—	Heavy	—[3]	0.03 to 0.10	0.0011	
Cylindrical roller, double row	Cylindrical	—	Heavy	—	—	0.0011	
Needle roller single row	Cylindrical	—	Heavy	—	—	0.0025	
Tapered roller single row	Cylindrical	—	Heavy[1]	Medium and heavy	—	0.0018	
Spherical roller double row	Cylindrical or tapered	—	Very heavy	Light and medium	1.5 to 3.0	0.0018	
Thrust ball single row	Cylindrical	—	—	Light and medium	—	0.0013	
Angular contact thrust ball double row	Cylindrical	—	—	Medium	—	0.0013	
Spherical roller thrust	Cylindrical	—	Not exceed 55% of simultaneously acting axial load	Heavy	1.5 to 3.0	Refer to manufacturer	

Information given in this chart is for general guidance only.

(1) **Must have simultaneously acting axial load or be mounted against opposed bearing.**
(2) **Must carry predominant axial load.**
(3) **Cylindrical roller bearings with flanges on inner and outer rings can carry axial loads providing the lubrication is adequate.**
(4) **The degree of misalignment permitted is dependent on internal design and manufacturers should be consulted.**

(5) The friction coefficients given in this table are approximate and will enable estimates to be made of friction torque in different types of bearings.

$$\text{Friction torque} = \mu P \frac{d}{2} 10^{-3} \text{ Nm}$$

Where P = bearing load, N
d = bearing bore diameter, mm
μ = friction coefficient.

Speed limits for radial bearings

Relatively high speeds in general applications can be achieved without resorting to special measures. The following information can be used as a guide to determining the speed limits for various types of bearings and assumes adequate care is taken with control of internal clearance, lubrication, etc. Figure 20.3 gives an approximate guide to limiting speeds and is based on loading giving a rating life of 100 000 hours. The curves are related to the expression

$$nd_m = f_1 f_2 A$$

where:

n = rotational speed, rev/min
d_m = mean diameter of bearing, $0.5 (D+d)$ mm
D = bearing outside diameter, mm
d = bearing bore diameter, mm
A = factor depending on bearing design (Tables 20.2 and 20.3)
f_1 = correction factor for bearing size (Fig. 20.1); this has been taken into account in the preparation of the curves (Fig. 20.3).
f_2 = correction factor for load (Fig. 20.2).
The rating life L_h is obtained from the nomogram.

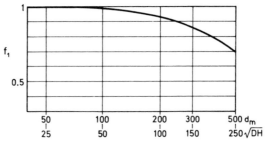

Fig. 20.1.

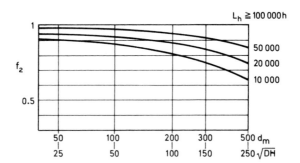

Fig. 20.2.

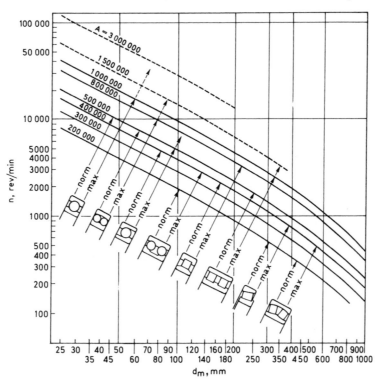

Fig. 20.3.

The limit curves, which are shown by a line of dashes, namely $A = 1\,500\,000$ and $A = 3\,000\,000$, indicate extreme values which have been achieved in a few specialist applications. The design, mounting and operation of such applications require considerable care and experience.

For spherical and tapered roller bearings the approximate speed limits apply to predominantly radial loads. Lower limits apply when the loads are predominantly axial.

When it is necessary to mount bearings such as angular contact single row ball bearings in pairs, the speed limit indicated by the curves is reduced by approximately 20%.

Table 20.2

Design of bearing	Factor A		Remarks
Single row deep groove ball bearings	Normal	500 000	Pressed steel cage
	Maximum	1 000 000	Solid cage
	Maximum	1 500 000	Solid cage, spray lubrication
Self-aligning double row ball bearings	Normal	500 000	Pressed steel cage
	Maximum	800 000	Solid cage
Angular contact single row ball bearings, $\alpha = 40$	Normal	400 000	Pressed steel cage
	Maximum	650 000	Solid cage, carefully controlled oil lubrication
$\alpha = 15$	Maximum	1 000 000	Solid cage
Angular contact double row ball bearings,	Normal	200 000	
	Maximum	400 000	C3 clearance (greater than normal)
Cylindrical roller single row bearings	Normal	400 000	Pressed steel cage
	Maximum	800 000	Solid cage
Cylindrical roller double row bearings	Normal	500 000	
	Maximum	1 000 000	Oil lubrication
	Maximum	1 300 000	Spray lubrication, maximum precision
Spherical roller bearings	Normal	200 000	For a predominantly axial load, limits of 150 000 (grease
	Maximum	400 000	lubrication) and 250 000 (oil lubrication) are used
Tapered roller single row bearings	Normal	200 000	For a predominantly axial load 20–40% lower values
	Maximum	400 000	should be assumed, depending on difficulty of working conditions

Table 20.3

Type of bearing	Factor A		Remarks
Angular contact thrust ball bearings	Normal	250 000	Grease lubrication
	Normal	300 000	Oil lubrication
	Maximum	400 000	Oil lubrication, cooling
Thrust ball bearings	Normal	100 000	
	Maximum	200 000	Solid cage
Spherical roller thrust bearings	Normal	200 000	Oil lubrication
			Good natural cooling usually sufficient
	Maximum	400 000	Effective cooling required

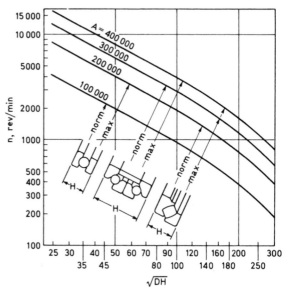

Fig. 20.4.

Speed limits for thrust bearings

An approximate guide to limiting speeds can be obtained from the curves in Fig. 20.4. These are based on the expression

$$n\sqrt{DH} = f_1 f_2 A,$$

where H equals the height of bearing in mm (for double row angular contact thrust ball bearings $H/2$ replaces H). All other factors are as for radial bearings.

Ball and roller thrust bearings for high-speed applications must always be preloaded or loaded with a minimum axial force and the bearing manufacturer should be consulted.

LUBRICATION

Grease lubrication

Grease lubrication is generally used when rolling bearings operate at normal speeds, loads and temperature conditions. For normal application the bearings and housings should be filled with grease up to 30–50% of the free space. Overpacking with grease will cause overheating. When selecting a grease, the consistency, rust-inhibiting properties and temperature range must be carefully considered. The relubrication period for a grease-lubricated bearing is related to the service life of the grease and can be estimated from the expression:

$$t_f = k \left(\frac{14 \times 10^6}{n\sqrt{d}} - 4\,d \right)$$

where:

t_f = service life of grease or relubrication interval, hours
k = factor dependent on the type of bearing (Table 20.4)
n = speed, rev/min
d = bearing bore diameter, mm

or from the curves given in Fig. 20.5.

Table 20.4

Bearing type	Factor k for calculation of re-lubrication interval
Spherical roller bearings, tapered roller bearings	1
Cylindrical roller bearings, needle roller bearings	5
Radial ball bearings	10

The amount of grease required for relubrication is obtained from:

$$G = 0.005\,DB$$

where:

G = weight of grease, g
D = bearing outside diameter, mm
B = bearing width, mm

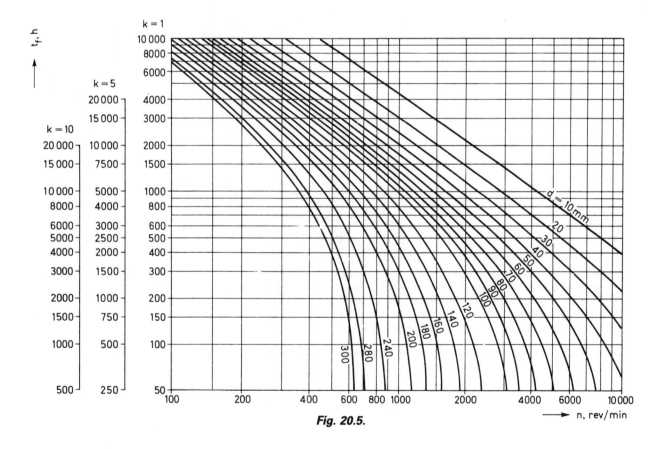

Fig. 20.5.

Oil lubrication

Oil lubrication is used when operating conditions such as speed or temperature preclude the use of grease. Fig. 20.6 gives a guide to suitable oil viscosities for rolling bearings taking into account the bearing size and operating temperature.

In the figure d = bearing bore diameter mm
                n = rotational speed rev/min
An example is given below and shown on the graph by means of the lines of dashes.

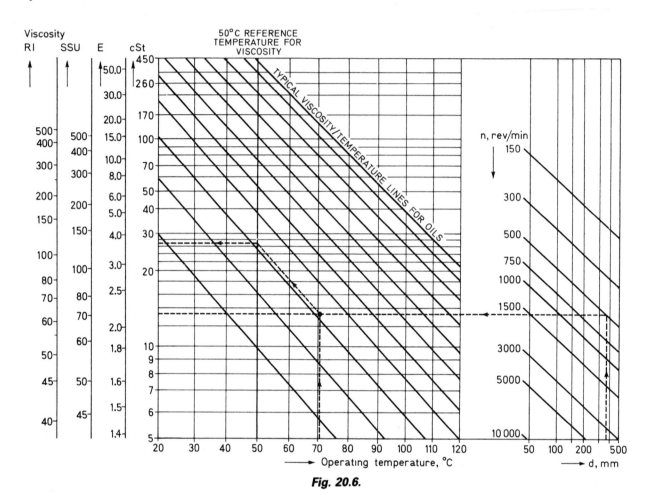

Fig. 20.6.

Generally the oil viscosity for medium and large size bearings should not be less than 12 centistokes at the operating temperature. For small high-speed bearings less viscous oils are used in order to keep friction to a minimum.

Example: A bearing having a bore diameter d = 340mm and operating at a speed n = 500 rev/min requires an oil having a viscosity of 13·2 centistokes at the operating temperature. If this operating temperature is assumed to be 70°C an oil having a viscosity of 26 centistokes at 50°C should be selected.

Acknowledgement is made to SKF (U.K.) Limited for permitting the use of graphical and tabular material.

COMPOSTION AND PROPERTIES

The stress levels in rolling bearings limit the choice of materials to those with a high yield and high creep strength. In addition, there are a number of other important considerations in the selection of suitable materials, including the following:

High impact strength
Good wear resistance
Corrosion resistance
Dimensional stability
High endurance under fatigue loading
Uniformity of structure

Steels have gained the widest acceptance as rolling contact materials as they represent the best compromise amongst the requirements and also because of economic considerations. The through hardening steels are listed in Table 21.1. The 535 grade is most popular (formerly EN31), but is not satisfactory at elevated temperatures due to loss of hardness and fatigue resistance. High-speed tool steels containing principally tungsten and molybdenum are superior to other materials at elevated temperatures, assuming adequate lubrication. Heat treatment is very important to give a satisfactory carbide structure of optimum hardness in order to achieve the maximum rolling contact fatigue life. Steelmaking practice can also influence results, e.g. refinements which improve mechanical properties and increase resistance to failure are vacuum melting and modification.

Table 21.1 Through hardening rolling-element bearing steels

(Increasing hardenability ↓)

Specification	*Maximum operating temperature* °C	*Composition %* C	Si	Mn	Cr	W	V	Mo	Ni	*Other elements*	*Notes*
SAE 50100*	120	0.95/1.10	0.20/0.35	0.25/0.45	0.40/0.60						Used for small thin parts, needle rollers etc.
SAE 51100	140	0.95/1.10	0.20/0.35	0.25/0.45	0.90/1.25						
534A99†	160	0.95/1.10	0.10/0.35	0.25/0.40	1.20/1.60						Most popular steels, used for majority of ball and roller bearings. Equivalent grades, previously SAE 52100 and EN31 respectively
535A99	160	0.95/1.10	0.10/0.35	0.40/0.70	1.20/1.60						
A‡	260	0.95/1.10	0.25/0.45	0.25/0.55	1.30/1.60					Al 0.75/1.25	Modified 535 A99 grades, generally by increase of Mn or Mo in heavy sections to give increase of hardenability
B	180	0.90/1.05	0.50/0.70	0.95/1.25	0.90/1.15						
C	180	0.85/1.00	0.60/0.80	1.40/1.70	1.40/1.70						
D	180	0.64/0.75	0.20/0.35	0.25/0.45	0.15/0.30			0.08/0.15	0.07/1.00		An alternative to 535A99, little used
5Cr5Mo	310	0.65	1.20	0.27	4.60		0.55	5.2			Intermediate high temperature steels. M50 most widely used
M50*	310	0.80	0.25	0.30	4.0		1.0	4.25			
BT1§	430	0.70	0.25	0.30	4.0	18.0	1.0				High hardenability steel used for large sections
M10	430	0.85	0.30	0.25	4.0		2.0	8.0			
BM1	450	0.80	0.30	0.30	4.0	1.50	1.0	8.0			
BM2	450	0.83	0.30	0.30	3.85	6.15	1.90	5.0			
4Cr7W2V4Mo5Co	540	1.07	0.02	0.30	4.40	6.80	2.0	3.90		Co 5.2	Superior hot hardness to the Mo base tool steels (660 VPN after 500 h at 540°C)

* AISI—SAE Steel Classification.
† BS 970 Pt II.
‡ Proprietary Steels, no specifications.
§ BS 4659.

Table 21.2 Physical characteristics of 535 grade (formerly EN31)

Young's Modulus	20°C	204 GN/m² (13 190 tsi)
	100°C	199 GN/m² (12 850 tsi)
Thermal expansion	20 to 150°C	13.25×10^{-6}/°C
Specific heat	20 to 200°C	0.107 cal/gm °C
Thermal conductivity	0.09 cgs	
Electrical resistivity	20°C	35.9 $\mu\Omega$/cm³
Density	20°C	7.7 gm/cc

Carburising is an effective method of case hardening, and is used for convenience in the manufacture of large rolling elements and the larger sizes of race. Conventional bearing steels pose through hardening and heat treatment difficulties in the larger sizes and sections. A deep case with the correct structure, supported on a satisfactory transition zone, can be as effective as a through hardened structure. Carburising steels are listed in Table 3, many of these being alternative to the through hardening steels used for normal bearing sizes. Other surface modifying techniques such as nitriding and carbide deposition are not effective.

Table 21.3 Carburising rolling-element bearing steels suitable for a maximum operating temperature of 180°C

	Specifica-tion	Composition %						Notes
		C	Si	% Mn	Cr	Ni	Mo	
↑ Increasing hardenability Increasing shock resistance Increasing core hardness ↓	SAE 1015	0.13/0.18		0.30/0.60				Little used. Mainly for small thin parts such as needle rollers etc.
	SAE 1019	0.15/0.20		0.70/1.00				
	SAE 1020	0.18/0.23		0.30/0.60				
	SAE 1024	0.19/0.25		1.35/1.65				
	SAE 1118	0.14/0.20		1.30/1.60				
	SAE 4023	0.20/0.25	0.20/0.35	0.70/0.90			0.20/0.30	
	SAE 4027	0.25/0.30	0.20/0.35	0.70/0.90			0.20/0.30	
	SAE 4422	0.20/0.25	0.20/0.35	0.70/0.90			0.35/0.45	
	SAE 5120	0.17/0.22	0.20/0.35	0.70/0.90	0.70/0.90			
	SAE 4118	0.18/0.23	0.20/0.35	0.70/0.90	0.40/0.60		0.08/0.15	
	805 M17*	0.14/0.20	0.10/0.35	0.60/0.95	0.35/0.65	0.35/0.75	0.15/0.25	
	805 M20	0.17/0.23	0.10/0.35	0.60/0.95	0.35/0.65	0.35/0.75	0.15/0.25	
	805 M22	0.19/0.2	0.10/0.35	0.60/0.95	0.35/0.65	0.35/0.75	0.15/0.25	
	665 M17	0.14/0.20	0.10/0.35	0.35/0.75		1.5/2.0	0.20/0.30	Equivalent grades. Most used
	665 M20	0.17/0.23	0.10/0.35	0.35/0.75		1.5/2.0	0.20/0.30	
	665 M23	0.20/0.26	0.10/0.35	0.35/0.75		1.5/2.0	0.20/0.30	
	SAE 4720	0.17/0.22	0.20/0.35	0.50/0.70	0.35/0.55	0.9/1.20	0.15/0.25	Modified 665 grades. Increased Cr and/or Ni contents increase hardenability on heavier sections
	SAE 4820	0.18/0.23	0.20/0.35	0.50/0.70		3.25/3.75	0.20/0.30	
	SAE 4320	0.17/0.22	0.20/0.35	0.45/0.65	0.40/0.60	1.65/2.0	0.20/0.38	
	SAE 9310	0.08/0.13	0.20/0.35	0.45/0.65	1.0/1.4	3.0/3.5	0.08/0.15	For very high shock resistance and core hardness, can increase nickel content to 4%
	SAE 3310	0.08/0.13		0.45/0.60	1.4/1.75	3.25/3.75		

* BS 970 Pt III

For use in a corrosive environment, martensitic stainless steels are used, Table 21.4. These generally have a poorer fatigue resistance than the through hardening steels at ambient temperatures, but are preferred for elevated temperature work.

In special applications, where the temperature or environment rules out the use of steels, refractory alloys, cermets and ceramics can be used. However, there are no standards laid down for the choice of very high temperature bearing materials and it is left to the potential user to evaluate his own material. Some refractory alloys which have been successfully used are listed in Table 21.5 and some ceramic materials in Table 21.6. Generally their use has not been widespread due to their limited application, high cost and processing difficulties, and with some their brittle nature.

Table 21.4 Stainless rolling-element bearing steels (martensitic)

Material	Maximum operating temperature °C	Composition %								Notes
		C	Si	Mn	Cr	W	V	Mo	Other	
440C*	180	1.03	0.41	0.48	17.3		0.14	0.50		Conventional material low hot hardness. Satisfactory for cryogenic applications
14Cr4Mo	430	0.95/1.20	<1.00	<1.00	13.0/16.0		<0.15	3.75/4.25		Modified 440C can be used as a high temperature bearing steel or corrosion resistant tool steel. Same workability as 440C
(a) 12Cr5Mo	430	0.70	1.00	0.30	12.0		—	5.25		Modified 14/4 grades. These have greater hardenability and increased hot hardness a→d as well as corrosion resistance
(b) 15Cr4Mo1V	430	1.15	0.30	0.50	14.5		1.10	4.0		
(c) 15Cr4Mo2V	430	1.20	0.30	0.50	14.5		2.0	4.0		
(d) 15Cr4Mo5Co	500	1.10/1.15	<0.15	<0.15	14.0/16.0	2.0/2.5	2.5/3.0	3.75/4.25	Co 5.0/5.5	

* AISI Steel Classification.

Table 21.5 Refractory alloy rolling-bearing materials

Material composition*	Maximum operating temperature °C	Composition %									Notes
		C	Cr	W	Ni	Fe	Co	Mo	Ti	Al	
Haynes 6B	540	1.1	30	4.5	3.0	3.0	Δ				Cobalt base, general proprietary compositions. Some with high carbon contents cannot be forged or machined so have to be cast and ground
Haynes 6K	to 310	1.6	31	—	—	—	Δ				
Stellite 100	,,	2	34	19	—	—	43				
Haynes 25 (Forging Stock)	,,	0.1	20	15	10	3	Δ				
Stellite 3	,,	2.4	30	13	—	—	52				
Rene 41	,,	0.09	19	—	Δ	—	11	10	3.1	1.5	Nickel base
TZM	Not known	99% Mo, 0.5% Ti, 0.08% Zr									Not for use in oxidising atmosphere at high temperature. Works well in liquid sodium or potassium

* Proprietary materials, no standard specifications. Δ means remaining % to 100%.

Table 21.6 Cermet and ceramic rolling-element bearing materials

Material	Maximum operating temperature °C	Composition	Notes
Alumina	above 800°C	99% Al_2O_3, hot pressed	
Titanium carbide cermet	,,	44% Ti; 4.5% Nb; 0.3% Ta; 11.0% C; 33% Ni; 7% Mo	Most successful in high temperature field
Tungsten carbide	,,	90/94% WC; 6.0/9.0% C	Brittle, oxidised at temperature. Works best with small carbide size and minimum of matrix material
Zirconium oxide	,,	Zr_2O_3	Little experience so far
Silicon nitride	,,	Si_3N_4 reaction bonded	Light loads, excellent resistance to thermal shock

MATERIAL SELECTION AND PERFORMANCE

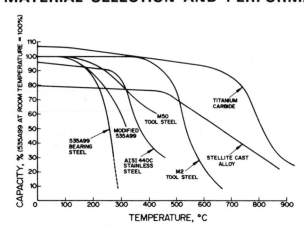

Fig. 21.1. Load capacity of rolling-element bearing materials

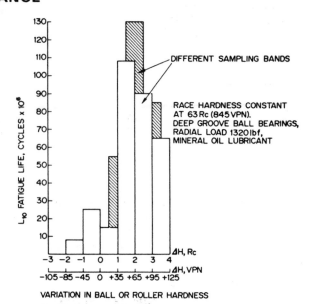

Fig. 21.2. Effect of hardness differences in 534A99 rolling-element bearing material

Rolling-element bearings are used where low friction, moderate loads and accurate positioning of a shaft are required. Because of the high contact stresses between the rolling elements and the races, precision can only be maintained at temperature with a high hardness. The effect of temperature on the load capacity of rolling contact bearings is shown in Fig. 21.1 where a significant drop in load limit can be seen above a temperature level characteristic for each material.

Through hardening 535A99 steel (EN31) is the most commonly used steel for conventional bearings. To ensure maximum life, the hardnesses of both races and rolling elements should be within an optimum hardness range, the balls or rollers being up to 10% harder than the races, Fig. 21.2. This corresponds to two points on the Rockwell C scale or 70–80 on the Vickers hardness scale when properly matched. However, current heat treatment procedures result in bearing components with significant hardness variations and bearings with a controlled hardness difference need to be specially ordered.

The combination of materials used in rolling contact seems to be important, the elastic and plastic material properties having a significant effect on the incidence of failure. For instance, a ceramic on ceramic combination will generally give a very short life because of the very high contact stresses produced, whereas ceramic rolling elements used in combination with steel races will give a considerably increased life. Information on some material combinations is given in Table 21.7.

Table 21.7 Roller / ball and race material combinations

Maximum operating temperature °C	Race material	Rolling element material	Notes
150	Carburised	535A99 or BTI	Used for the smaller size of rolling element
—	M10	535A99	Not recommended. Reduced rolling contact fatigue lives below that of the tool steel— tool steel combination
—	BTI	535A99	
—	M50	535A99	
480	BM2	TiC	Performance equivalent to 535A99 on 535A99 at room temperature
Up to 480	14/4	TiC	
650	Chromium carbide	Stellite 100	Limited life
820	TiC Cermet	Al$_2$O$_3$	Limited life

SELECTION OF CAGE MATERIALS

The most commonly used materials are steel, brass, bronze and plastics. Steel retainers are generally manufactured from riveted strips, while bronze and plastic cages are usually machined.

There is a number of important considerations in the selection of materials, including the following:

Resistance to wear   To withstand the effect of sliding against hardened steel
Strength   To enable thin sections to be used
Resistance to environment   To avoid corrosion, etc.
Suitability for production   Usually machined or fabricated

Table 21.8 Typical materials and their limitations

Material	Temperature °C	Wear resistance	Oxidation resistance	Remarks
Low carbon steel	260	Fair	Poor	Standard material for low-speed or non-critical applications
Iron silicon bronze	320	Good 150°C Excellent 260°C	Excellent	Jet engine applications as well as other medium-speed, medium-temperature bearings
S Monel	535	Fair	Excellent	Excellent high temperature strength
AISI 430 stainless steel	535	Poor	Excellent	Standard material for 440C stainless steel bearings—low speed
17–4–Ph stainless steel	535	Poor in air	Excellent	Good high temperature performance. Good wear resistance
Non-metallic retainers, fabric base phenolic laminates	135	Excellent	—	High-speed bearing applications
Silver plate	Possibly 180	Excellent to 150°C	—	Has been used in applications where marginal lubrication was encountered during part of the operating cycle

SHAFT AND HOUSING DESIGN

Rigidity

1   Check the shaft slope at the bearing positions due to load deflection, unless aligning-type bearings are to be used.
2   Check that the housing gives adequate support to the bearing outer ring, and that housing distortion under load will not cause distortion of the bearing outer ring.
3   Design the housing so that the resultant bearing slope is subtractive — see Figs. 22.1(a) and 22.1(b).

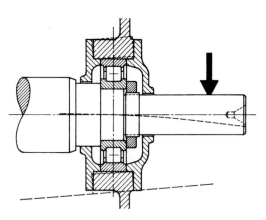

Fig. 22.1(a). Incorrect — slopes adding

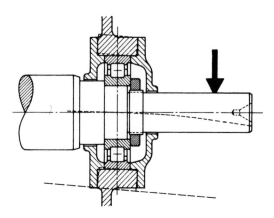

Fig. 22.1(b). Correct — slopes subtracting

Alignment

1   For rigid-type bearings, calculate the shaft and housing slopes due to load deflection.
2   Determine the errors of housing misalignment due to tolerance build-up.
3   Ensure that the total misalignment does not exceed the values given in Table 22.1.

Table 22.1 Approximate maximum misalignments for rigid bearings

Rigid bearing type	Permitted misalignment
Radial ball bearings	1.0 mrad
Angular contact ball bearings	0.4 mrad
Radial roller bearings	0.4 mrad
Needle roller bearings	0.1 mrad

Seatings

1 The fits indicated in Table 22.2 should be used to avoid load-induced creep of the bearing rings on their seatings.

Table 22.2 Selection of seating fit

Rotating member	Radial load	Shaft seating	Housing seating
Shaft	Constant direction	Interference fit	Sliding or transition fit
Shaft	Rotating	Clearance fit	Interference fit
Shaft *or* housing	Combined constant direction and rotating	Interference fit	Interference fit
Housing	Constant direction	Clearance fit	Interference fit
Housing	Rotating	Interference fit	Sliding or transition fit

2 Bearings taking purely axial loads may be made a sliding fit on both rings as there is no applied creep-inducing load.

3 Select the shaft and housing seating limits from Tables 22.3 and 22.4, respectively, these having been established to suit the external dimensions, and internal clearances, of standard metric series bearings.

4 Where a free sliding fit is required to allow for differential expansion of the shaft and housing use H7.

Table 22.3 Shaft seating limits for metric bearings (values in micro-metres)

Shaft mm	over	—	6	10	18	30	50	80	120	150	180	250	315
	incl.	6	10	18	30	50	80	120	150	180	250	315	400
Int. fit	grade	j5	j5	j5	j5	j5	k5	k5	k5	m5	m5	n6	n6
	limits	+3 −2	+4 −2	+5 −3	+5 −4	+6 −5	+15 +2	+18 +3	+21 +3	+33 +15	+37 +17	+66 +34	+73 +37
Sliding fit	grade	g6	g6	g6	g6	g6	g6	g6	g6	g6	g6	g6	g6
	limits	−4 −12	−5 −14	−6 −17	−7 −20	−9 −25	−10 −29	−12 −34	−14 −39	−14 −39	−15 −44	−17 −49	−18 −54

EXAMPLE

Interference fit shaft 35 mm dia. tolerance from table = +6/−5µm. Therefore, shaft limit = 35.006/34.995 mm.

Table 22.4 Housing seating limits for metric bearings (values in micro-metres)

Hsg mm	over	—	6	10	18	30	50	80	120	180	250	315	400	500	630
	incl.	6	10	18	30	50	80	120	180	250	315	400	500	630	800
Int. fit	grade	M6	M6	M6	M6	M6	M6	M6	M6	M6	M6	M6	M6	M6	M6
	limits	−1 −9	−3 −12	−4 −15	−4 −17	−4 −20	−5 −24	−6 −28	−8 −33	−8 −37	−9 −41	−10 −46	−10 −50	−26 −70	−30 −80
Transition fit	grade	J6	J6	J6	J6	J6	J6	J6	J6	J6	J6	J6	J6	H6	H6
	limits	+5 −3	+5 −4	+6 −5	+8 −5	+10 −6	+13 −6	+16 −6	+18 −7	+22 −7	+25 −7	+29 −7	+33 −7	+44 −0	+50 −0

EXAMPLE

Transition fit housing 72 mm dia. tolerance from table = +13/−6µm. Therefore, housing limit = 72.013/71.994 mm.

Seatings (continued)

4   Control the tolerances for out-of-round and conicity errors for the bearing seatings. These errors in total should not exceed the seating dimensional tolerances selected from Tables 22.3 and 22.4.

5   Adjust the seating limits if necessary, to allow for thermal expansion differences, if special materials other than steel or cast iron are involved. Allow for the normal fit at the operating temperature, but check that the bearing is neither excessively tight nor too slack at both extremes of temperature. Steel liners, or liners having an intermediate coefficient of thermal expansion, will ease this problem. They should be of at least equivalent section to that of the bearing outer ring.

6   Avoid split housings where possible. Split housings must be accurately dowelled before machining the bearing seatings, and the dowels arranged to avoid the two halves being fitted more than one way round.

Abutments

1   Ensure that these are sufficiently deep to provide adequate axial support to the bearing faces, particularly where axial loads are involved.

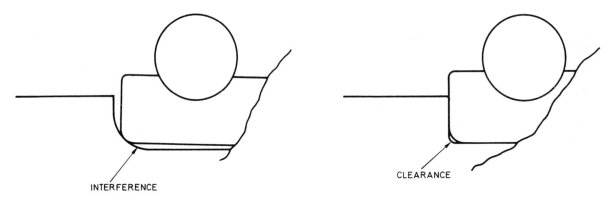

INTERFERENCE

Fig. 22.2(a). Incorrect

CLEARANCE

Fig. 22.2(b). Correct

2   Check that the seating fillet radius is small enough to clear the bearing radius — see Figs. 22.2(a) and 22.2(b). Values for maximum fillet radii are given in the bearing manufacturers' catalogues and in ISO 582 (1979).

3   Design suitable grooves into the abutments if bearing extraction is likely to be a problem.

BEARING MOUNTINGS

Horizontal shaft

1 The basic methods of mounting illustrated in Figs. 22.3(a) and 22.3(b) are designed to suit a variety of load and rotation conditions. Use the principles outlined and adapt these mountings to suit your particular requirements.
2 The type of mounting may be governed more by end-float or thermal-expansion requirements than considerations of loading and rotation.

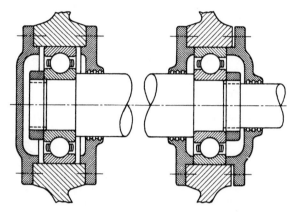

Fig. 22.3(a). Two deep groove radial ball bearings

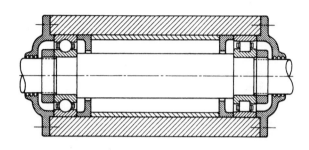

Fig. 22.3(b). One ball bearing with one cylindrical roller bearing

Condition	Suitability
Rotating shaft	Yes
Rotating housing	No
Constant direction load	Yes
Rotating load	No
Radial loads	Moderate capacity
Axial loads	Moderate capacity
End-float control	Moderate
Relative thermal expansion	Moderate

Condition	Suitability
Rotating shaft	Yes
Rotating housing	Yes
Constant direction load	Yes
Rotating load	Yes
Radial loads	
Location bearing	Moderate capacity
Non-location bearing	Good capacity
Axial loads	Moderate capacity
End-float control	Moderate
Relative thermal expansion	Yes

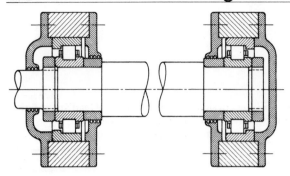

Fig. 22.3(c). Two lip-locating roller bearings

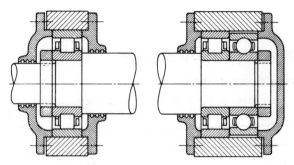

Fig. 23.3(d). Two roller bearings with 'loc' location pattern ball bearing which has reduced o.d. so that it does not take radial loads

Condition	Suitability
Rotating shaft	Yes
Rotating housing	No
Constant direction load	Yes
Rotating load	No
Radial loads	Good capacity
Axial loads	Low capacity
End-float control Relative thermal expansion	Sufficient end float required to allow for tolerances and temperature

Condition	Suitability
Rotating shaft	Yes
Rotating housing	Yes
Constant direction load	Yes
Rotating load	Yes
Radial loads	Good capacity
Axial loads	Moderate capacity
End-float control	Moderate
Relative thermal expansion	Yes

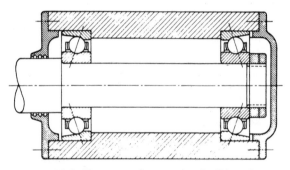

Fig. 22.3(e). Two angular contact ball bearings

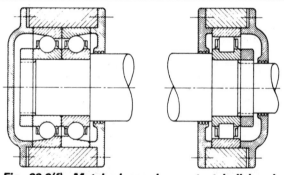

Fig. 22.3(f). Matched angular contact ball bearing unit with roller bearing

Condition	Suitability
Rotating shaft	No
Rotating housing	Yes
Constant direction load	Yes
Rotating load	No
Radial loads	Moderate capacity
Axial loads	Good capacity
End-float control	Good
Relative thermal expansion	Allow for this in the initial adjustment

Condition	Suitability
Rotating shaft	Yes
Rotating housing	Yes
Constant direction load	Yes
Rotating load	Yes
Radial loads	Good capacity
Axial loads	Good capacity
End-float control	Good
Relative thermal expansion	Yes

Vertical shaft

1 Use the same principles of mounting as indicated for horizontal shafts.
2 Where possible, locate the shaft at the upper bearing position because greater stability is obtained by supporting a rotating mass at a point above its centre of gravity.
3 Take care to ensure correct lubrication and provide adequate means for lubricant retention. Use a No. 3 consistency grease and minimise the space above the bearings to avoid slumping.
4 Figure 22.4 shows a typical vertical mounting for heavily loaded conditions using thrower-type closures to prevent escape of grease from the housings.

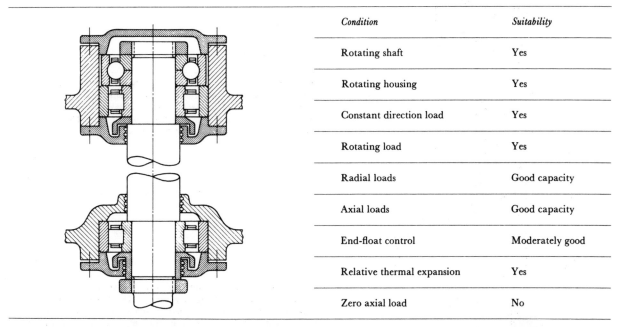

Condition	Suitability
Rotating shaft	Yes
Rotating housing	Yes
Constant direction load	Yes
Rotating load	Yes
Radial loads	Good capacity
Axial loads	Good capacity
End-float control	Moderately good
Relative thermal expansion	Yes
Zero axial load	No

Fig. 22.4. Vertical mounting for two roller bearings and one duplex location pattern bearing, which has reduced o.d. so that it does not take radial loads

5 For high speeds use a stationary baffle where two bearings are used close together. This will minimise the danger of all the grease slumping into the lower bearing (Fig. 22.5).

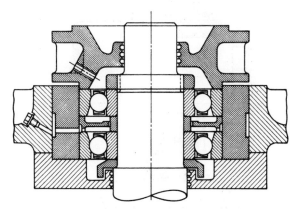

Fig. 22.5. Matched angular contact unit with baffle spacer

Fixing methods

Type of fixing	Description
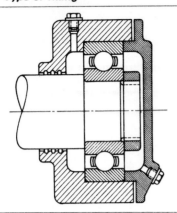	Shaft—screwed nut provides positive clamping for the bearing inner ring Housing—the end cover should be spigoted in the housing bore, *not* on the bearing o.d., and bolted up uniformly to positively clamp the bearing outer ring squarely
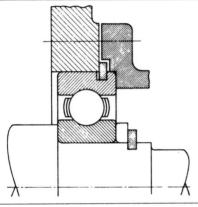	Circlip location can reduce cost and assembly time. Shaft—use a spacer if necessary to provide a suitable abutment. Circlips should not be used if heavy axial loads are to be taken or if positive clamping is required (e.g. paired angular contact unit). Housing shows mounting for snap ring type of bearing
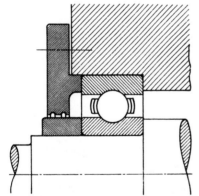	Interference fit rings are sometimes used as a cheap and effective method of locating a bearing ring axially. The degree of interference must be sufficient to avoid movement under the axial loads that apply. Where cross-location is employed, the bearing seating interference may give sufficient axial location
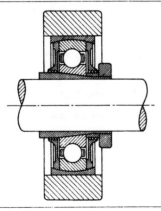	Bearing with tapered clamping sleeve. This provides a means for locking a bearing to a parallel shaft. The split tapered sleeve contracts on to the shaft when it is drawn through the mating taper in the bearing bore by rotation of the screwed locking nut

Fig. 22.6. Methods of fixing bearing rings

Sealing arrangements

1 Ensure that lubricant is adequately retained and that the bearings are suitably protected from the ingress of dirt, dust, moisture and any other harmful substances. Figure 22.7 gives typical sealing methods to suit a variety of conditions.

Sealing arrangement	*Description*
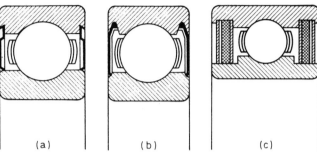 (a) (b) (c)	(a) Shielded bearing—metal shields have running clearance on bearing inner ring. Shields non-detachable; bearing 'sealed for life' (b) Sealed bearing—synthetic rubber seals give rubbing contact on bearing inner ring, and therefore improved sealing against the ingress of foreign matter. Sealed for life (c) Felt sealed bearing—gives good protection in extremely dirty conditions
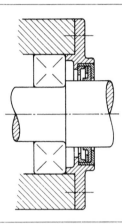	Proprietary brand rubbing seals are commonly used where oil is required to be retained, or where liquids have to be prevented from entering the bearing housing. Attention must be given to lubrication of the seal, and the surface finish of the rubbing surface
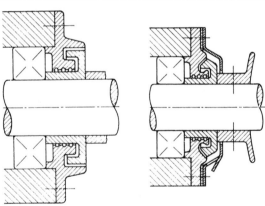	Labyrinth closures of varying degrees of complexity can be designed to exclude dirt and dust, and splashing water. The diagram shown on the left is suitable for dusty atmospheres, the one on the right has a splash guard and thrower to prevent water ingress. The running clearances should be in the region of 0.2 mm and the gap filled with a stiff grease to improve the seal effectiveness

Fig. 22.7. Methods of sealing bearing housing

BEARING FITTING

1 Ensure cleanliness of all components and working areas in order to avoid contamination of the bearings and damage to the highly finished tracks and rolling elements.
2 Check that the bearing seatings are to the design specification, and that the correct bearings and grades of clearance are used.
3 Never impose axial load through the rolling elements when pressing a bearing on to its seating—apply pressure through the race that is being fitted — see Figs. 22.8(a) and 22.8(b). The same principles apply when extracting a bearing from its seatings.

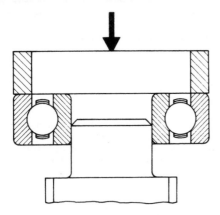

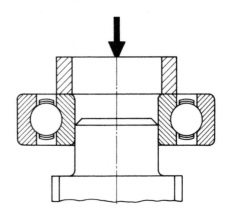

Fig. 22.8(a). Incorrect — load applied through outer ring when fitting inner ring

Fig. 22.8(b). Correct — load applied through ring being fitted

4 When shrink-fitting bearings on their seatings, never heat the bearings above 120°C and always ensure the bearing is firmly against its abutment when it has cooled down.
5 Where bearing adjustment has to be carried out, ensure that the bearings are not excessively preloaded against each other. Ideally, angular contact bearings should have just a small amount of preload in the operating conditions, so it is sometimes necessary to start off with a degree of end float to allow for relative thermal expansion.
6 Ensure that the bearings are correctly lubricated. Too much lubricant causes churning, overheating and rapid oxidation and loss of lubricant effectiveness. Too little lubricant in the bearing will cause premature failure due to dryness.

Table 23.1 The selection of the type of slideway by comparative performance

	Plain		Rolling element		Hydrostatic	
	Metal/metal	Plastic/metal	Non recirculating	Recirculating	Liquid	Gas
Stroke	Any	Any	Short	Any	Any[1]	Any[1]
Lubricant	Oil, grease, use transverse oil grooves	Oil, grease, dry	Oil (oil mist), grease	Oil (oil mist), grease	Any (non corrosive)	Air (clean and dry)
Load capacity	Medium[2]	Medium, high at low speed[2]	Medium (consult maker)	Medium, high (consult maker)	Can be very high[3]	Medium low[4]
Speed	Medium, high[6]	Medium	Any (consult maker)	Any (consult maker)	Low, medium[5]	Any
Typical friction	See note 6					
Stiffness	High	High[7]	High	High	High (keep h small)[8]	Low, medium (keep h very small)[8]
Transverse damping	Good	Good	Low, medium varies with preload	Low, medium varies with preload	Good	Low, medium
Accuracy of linear motion	Good if ways ground or scraped	Good, beware variation in thickness of adhesive, etc.	Virtually that of guideway	Virtually that of guideway	Excellent, averages local geometrical errors.[9] May run warm	Excellent, averages local geometrical errors.[9] Runs cool
Materials	Any good bearing combination	Metal is usually CI or steel with finish better than 0.25μm cla[9]	Hardened steel (R_c60) guideways (proprietary insert strips available)	Hardened steel (R_c60) guideways (proprietary insert strips available)	Any[10]	Any[10]

Table 23.1—continued

	Plain		Rolling element		Hydrostatic	
	Metal/metal	*Plastic/metal*	*Non recirculating*	*Recirculating*	*Liquid*	*Gas*
Wear rate	Low/medium[11]	Low/medium[11]	Low[11]	Low[11]	Virtually none	Virtually none
Installation	Easy	Moderate	Moderate	Moderate	Requires pump, etc.	Requires air supply, etc.
Preload used (on opposed faces)	Negligible: it increases frictional resistance	Negligible: it increases frictional resistance	Needed to eliminate backlash, excess reduces life	Needed to eliminate backlash, excess reduces life	Inherent high, can distort a weak structure	Inherent
Protection required	Wipers, covers	Wipers,[12] covers	Wipers, covers	Wipers, covers	Covers, filter fluid for re-use	Wipers[12]
Initial cost	Low	Low, medium	Medium	Medium, high[13]	Medium, high[14]	Medium, high

(1) Fluid, at relatively high pressure is supplied to the *shorter* of the sliding members.
(2) Typically 50 to 500 kN m^{-2} for machine tools, otherwise use PV value for the material pair for boundary lubrication of a collar-type thrust bearing.
(3) Ultimate, typically 0.5 × supply pressure × area; working $\simeq$ 0.25 to 0.5 × ultimate.
(4) Limited, often, by air line pressure and area available.
(5) Prevent air entrainment by flooding the leading edge if the slide velocity exceeds the fluid velocity in the direction of sliding.
(6) Liable to stick-slip at velocities below 1 mm s^{-1}, use slideway oil with polar additive, stiffen the drive so that

$$[\text{drive stiffness (N m}^{-1})/\text{driven mass (kg)}]^{\ddagger} > 300$$

(7) Provided plastic facing or insert is in full contact with backing.
(8) $h > 3 \times$ geometrical error of bearing surfaces.
(9) Some sintered and PTFE impregnated materials must not be scraped or ground. Some resins may be cast, with high accuracy, against an opposing member (or against a master) and need no further finishing.
(10) Use a good bearing combination in case of fluid supply failure or overload. Consider a cast resin – see note 9.
(11) May be excessive if abrasive or swarf is present.
(12) Wiper may have to operate dry.
(13) Cost rises rapidly with size.
(14) Cost rises rapidly with size but more slowly than for rolling element bearings; may share hydraulic supplies.

Combined bearings

Hydrostatic (liquid) bearings are usually controlled by a restrictor (as illustrated) or by using constant flow pumps, one dedicated to each pocket.

Hydrostatic bearing style pockets supplied at constant pressure can be combined with a plain bearing to give a 'pressure assist', i.e., 'load relief', feature whilst still retaining the high stiffness characteristic of a plain bearing; a combination useful for cases of heavy dead-weight loading.

Hydrostatic bearing style lands, supplied via small pockets at a usually low constant pressure can be fitted around, or adjacent to, rolling element bearings to give improved damping in the transverse direction, used rarely and only when vibration mode shape is suitable.

Table 23.2 Notes on the layout of slideways
(generally applicable to all types shown in Table 23.1)

Geometry	Surfaces	Notes
SINGLE SIDED		This basic single-sided slide relies largely on mass of sliding member to resist lifting forces, plain slides tend to rise at speed, hydrostatics soft under light load. Never used alone but as part of a more complex arrangement
	3	Direction of net load limited, needs accurate V angle, usually plain slide
	4	Easy to machine; the double-sided guide slide needs adjustment, e.g. taper gib, better if b is small in relation to length
	4	Used for intermittent movement, often clamped when stationary, usually plain slide
	3 or 4	Accurate location, 3 ball support for instruments (2 balls in the double vee, 1 ball in the vee-flat)
DOUBLE SIDED		Resists loads in both directions
	4	Generally plain, adjusted by parallel gib and set screws, very compact
	6	All types, h large if separate thick pads used, make t sufficiently large so as to prevent the structure deforming, watch for relative thermal expansion across b if b is large relative to the clearance

Table 23.2 — *continued*

Geometry	Surfaces	Notes
	4	Usually plain or hydrostatic; watch thickness *t*. If hydrostatic an offset vertical load causes horizontal deflection also
	4	Ball bearings usually but not always, non recirculating, crossed-axis rollers are also used instead of balls Proprietary ball and roller, recirculating and non-recirculating units of many types involving both 2- and 4-track assemblies, complete with rails, are also available
		Rod guides must be in full contact with main body: (rod guides are shown cross hatched)
		Most types (including plain, hydrostatic, ball bushes or hour-glass-shaped rollers) bars liable to bend, bar centres critical, gaps or preload adjustment not easy
		Not usually hydrostatic, bars supported but might rock, bearings weaker, clearance adjustment is easy by 'springing' the slotted housing
		Plain or roller, torsional stiffness determined by bar usually
		Recirculating ball usually medium torque loads permitted except when plain bars are used

MATERIALS

Usual combinations	*Special precautions*	

The most usual combination is that of a steel pivot and a synthetic sapphire jewel. The steel must be of high quality, hardened and tempered, with the tip highly polished. The jewel also must be highly polished. Diamond jewels are sometimes used for very heavy moving systems. A slight trace of a good quality lubricating oil such as clock oil or one of the special oils made for this purpose, improves the performance considerably

The sapphire crystal has natural cleavage planes, and the optic axis, i.e. the line along which a ray of light can pass without diffraction, is at right angles to these planes. The angle between this optic axis and the line of application of the load is called the optic axial angle α, and experiment has conclusively demonstrated that, for the best results as regards friction, wear etc., this angle should be 90 degrees, and any departure from this produces a deterioration in performance

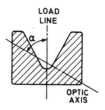

OPERATING CONDITIONS

Arrangement	*Remarks*

Vertical shaft

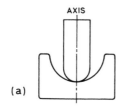

(a)

The pivot is cylindrical with a spherical end. The jewel is a spherical cup. This is used, for example, in compasses and electrical integrating metres. The optical axial angle can be controlled in this case

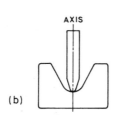

(b)

The pivot is cylindrical ending in a cone with a hemispherical tip. The jewel recess is also conical with a hemispherical cup at the bottom of the recess. This is used in many forms of indicating instrument and again the optic axial angle can be controlled

Horizontal shaft

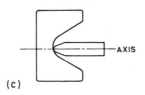

(c)

The pivot and jewel are the same as for the vertical shaft. In this case the optic axial angle cannot be controlled since the jewel is usually rotated for adjustment, so that the load on the jewels must be reduced in this case

PERFORMANCE CHARACTERISTICS

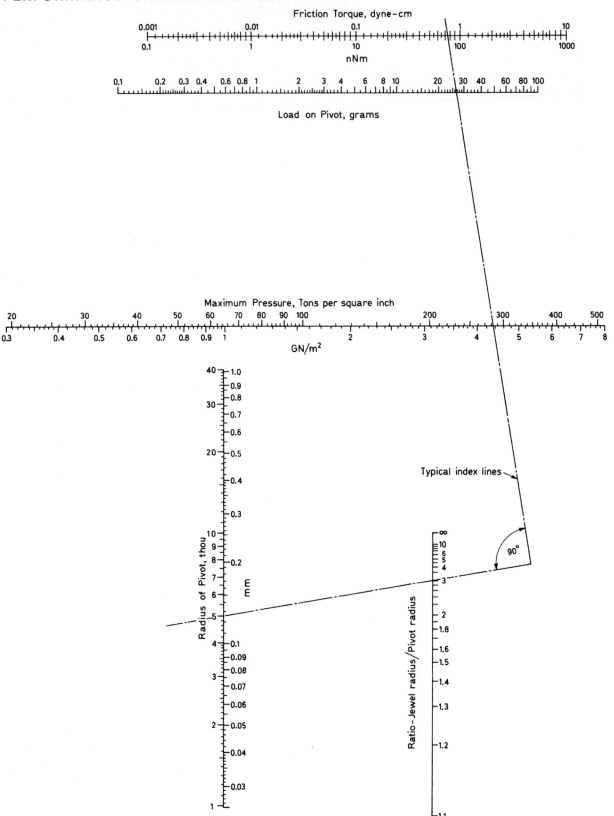

* The friction torque in the nomogram is calculated for a coefficient of friction of 0.1. For a dry jewel and pivot multiply friction torque by 1.6 and for an oiled combination by 1.4.

DESIGN

There are two important quantities which must be considered in designing a jewel/pivot system and in assessing its performance. These are the maximum pressure exerted between the surfaces of the jewel and pivot, and the friction torque between them. These depend on the dimensions and the elastic constants of the two components and can be determined by the use of the nomogram. This is of the set-square index type, one index line passes through the values of the pivot radius and the ratio jewel radius/pivot radius.

The second index line, at right angles to the first, passes through the value of the load on the jewel, and will then also pass through the values of maximum pressure and friction torque. The example shown is where the pivot radius is 5 thou. of an inch (0.127 mm), the ratio jewel radius/pivot radius is 3, and the load on the pivot is 27 grams. The resulting pressure is 282 tons per square inch (4.32 GN/m^2), and the friction torque 0.77 dyne-cm (77 nNm).

Loading	Remarks
Static	It is generally considered that the crushing strength of steel is about 500 tons per square inch, and experiment has shown that the sapphire surface cannot sustain pressures much above this without damage. If a safety factor of 2 is introduced then the maximum pressure should not exceed 250 tons per square inch. Unfortunately, an alteration in jewel and pivot design aimed at reducing the pressure, results in an increase in friction torque and vice versa, so that a compromise is usually necessary
Impact	All calculations have been based on static load on the jewel. Impact due to setting an instrument down on the bench, transport etc., can increase the pressure between jewel and pivot very considerably, and in many cases the jewel is mounted with a spring loading, so as to reduce the maximum force exerted on it. In general, the force required to move the jewel against the spring should not be more than twice the static load of the moving system. This spring force must then be taken as the load on the pivot

MATERIALS FOR FLEXURE HINGES AND TORSION SUSPENSIONS

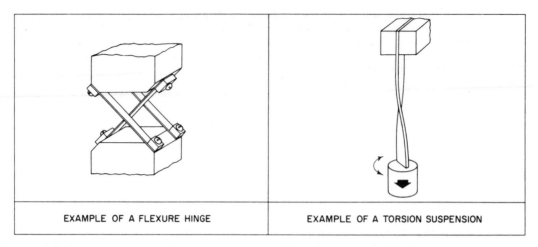

| EXAMPLE OF A FLEXURE HINGE | EXAMPLE OF A TORSION SUSPENSION |

Flexure hinges and torsion suspensions are devices which connect or transmit load between two components while allowing limited relative movement between them by deflecting elastically.

Selection of the most suitable material from which to make the elastic member will depend on the various requirements of the application and their relative importance. Common application requirements and the corresponding desired properties of the elastic member are listed in Table 25.1.

Table 25.1 Important material properties for various applications of flexure hinges and torsion suspensions

Application requirement	Desired material property
1. Small size	High maximum permissible stress, $f_{max} =$ yield strength, f_Y *unless* the application involves a sufficiently large number of stress cycles for fatigue to be the critical condition, in which case: $f_{max} =$ fatigue strength, f_F
2. Flexure hinge with maximum movement for a given size	High f_{max}/E: $E =$ Young's Modulus
3. Flexure hinge with the maximum load capacity for a given size and movement	High f_{max}^3/E^2
4. Flexure hinge with minimum stiffness (for a given pivot geometry)	High $1/E$: note that stiffness can be made zero or negative by suitable pivot geometry design
5. Torsion suspension with minimum stiffness for a given suspended load	High $f_{max}^2 \times U/G$: $G =$ shear modulus, $U =$ aspect ratio (width/thickness) of suspension cross-section. U is not a material property but emphasises the value of being able to manufacture the suspension material as thin flat strip

Application requirement	Desired material property
6. Elastic component has to carry an electric current	High electrical conductivity, k_e
7. Elastic component has to provide a heat path	High thermal conductivity, k_t
8. Elastic component has to provide the main reactive force in a sensitive measurement or control system	Negligible hysteresis and elastic after-effect; non magnetic
9. As 8 and may be subject to temperature fluctuations	Low temperature coefficient of thermal expansion and elastic modulus (E or G)
10. As 6 but current has to be measured accurately by system of which elastic component is a part	Low thermoelectric e.m.f. against copper (or other circuit conductor) and low temperature coefficient of electrical conductivity
11. Elastic component has to operate at high or low temperature	As for 1–10 above, but properties, for example strength, must be those at the operating temperature
12. Elastic component has to operate in a potentially corrosive environment (includes 'normal' atmospheres)	Appropriate, good, corrosion resistance, especially if requirements 8 or 10 have to be met

Table 25.2 Relevant properties of some flexure materials

Material	Yield strength[1] f_y N/m²×10⁷	Yield strength[1] f_y lbf/in²×10³	Fatigue strength[2] f_t N/m²×10⁷	Fatigue strength[2] f_t lbf/in²×10³	Young's Modulus E (For G see note 7) N/m²×10¹⁰	Young's Modulus E lbf/in²×10⁶	Thermal conductivity k_t W/m °C	Thermal conductivity k_t Btu/h ft °F	Electrical conductivity k_e %IACS[3]	Atmospheric corrosion resistance[4]	Approximate maximum continuous operating temperature in air °C	°F
Spring steels 0.6–1.0C 0.3–0.9Mn	80–210	120–300	40–70	60–100	21	30	45	26	9.5	P	230	450
Carbon chromium stainless steel (BS 420 S45)	150	200	60	85	21	30	24	14	2.8	M	540	1000
High strength alloy steels: nickel maraging steel	210	300	66	96	19	27	17	10	4	P	480	900
DTD 5192 (NCMV)	210	300	80	115	21	30	35	20	6	P	400	750
Inconel X	165	240	65	95	21	31	12	7	1.7	E	650	1200
High strength titanium alloy	95	140	65	95	11	16	9	5	1.1	G	480	900
High strength aluminium alloy	50	73	15	22	7.2	10.4	120	70	30	P	200	400
Beryllium copper	90	135	38	55	12.5	18	100	60	25	G	230	450
Low beryllium copper	65	95	24	35	11.5	16.5	170	100	45	G	200	400
Phosphor bronze (8% Sn; hard)	60	90	20	29	11	16	55	32	12	G	180	350
Glass fibre reinforced nylon (40% G.F.)	20	30	NA	NA	1.2	1.8	0.35	0.20	negligible	E	110	230
Polypropylene	3.7	5.4[5]	NA	NA	0.14	0.2[5]	0.17	0.1	negligible	E[6]	50	120

Notes: 1. Very dependent on heat treatment and degree of working. Figures given are typical of fully heat treated and processed strip material of about 0.1 in thickness at room temperature. Thinner strip and wire products can have higher yield strengths.

2. Fatigue strengths are typical for reversed bending of smooth finished specimens subjected to 10⁷ cycles. Fatigue strengths are reduced by poor surface finish and corrosion, and may continue to fall with increased cycles above 10⁷.

3. Percentage of the conductivity of annealed high-purity copper at 20°C.

4. Order of resistance on following scale: P—poor, M—moderate, G—good, E—excellent. Note, however, that protection from corrosion can often be given to materials which are poor in this respect by grease or surface treatments.

5. At high strain rate. Substantial creep occurs at much reduced stress levels, probably restricting applications to where the steady load is zero or very small, and the deflections are of short duration.

6. But the material deteriorates rapidly in direct sunlight.

7. Modulus of Rigidity, $G = E/2(1 + \text{Poisson's ratio}, v)$. For most materials $v \simeq 0.3$, for which $G \simeq E/2.6$.

NA Data not available.

The main properties of interest for selection of materials for flexures and torsion suspensions are given in Table 25.2 and Table 25.3. Values given are intended to provide a comparison of different materials, but they are only typical and should not be used to specify minimum properties.

Table 25.3　Some materials used to meet special requirements in accurate instruments or control systems

Requirement	Material(s)
Minimum hysteresis and elastic after-effect	91.5% platinum, 8.5% nickel alloy. Platinum/silver alloy, 85/15 to 80/20. Quartz
Zero thermoelectric e.m.f. against copper	Copper
Maximum corrosion resistance (instrument torsion suspensions)	Gold or platinum alloys, quartz
Maximum electrical conductivity	Silver
Torsionless suspensions (e.g. for magnetometers)	Stranded silk and other textile fibres)

MATERIALS FOR KNIFE EDGES AND PIVOTS

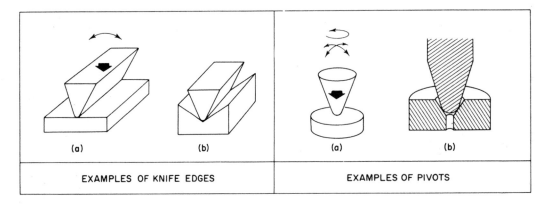

(a)	(b)	(a)	(b)
EXAMPLES OF KNIFE EDGES		EXAMPLES OF PIVOTS	

Knife edges and pivots are bearings in which two members are loaded together in nominal line or point contact respectively, and can tilt relative to one another through a limited angle by rotation about the contact; a pivot can also rotate freely about the load axis.

The main requirement of the materials for this type of bearing is high hardness, so that high load capacity can be provided, while keeping the width of the contact area small for low friction torque and high positional accuracy of the load axis.

Table 25.4 Important material properties for various applications of knife edges and pivots

Application requirement	Desired material property
1. High load capacity for a given bearing geometry	High $\dfrac{hardness^2}{modulus\ of\ elasticity}$
2. Ability to tolerate overload, impact or rough treatment generally	A measure of ductility in compression, so that overload can be accommodated by plastic deformation rather than chipping or fracture
3. Requirements 1 and 2 together (for example for weigh-bridges, strength-testing machines, etc.)	High hardness together with some ductility. In practice various metallic materials with hardnesses greater than 60 Rc (690 Knoop) are usually specified
4. Very low friction with useful load capacity where freedom from impact and overloading can be expected (for example in sensitive force balances and other delicate equipment)	Very high $\dfrac{hardness^2}{modulus\ of\ elasticity}$ using various brittle materials having exceptionally high hardness
5. High wear resistance	High hardness is generally beneficial
6. Little indentation of block by knife edge or pivot	Hardness of block > hardness of knife edge or pivot. (This is nearly always desirable; the differential should be at least 5%)
7. The two members of the bearing have to slide relative to one another at the contact (see examples (b)) and must be metallic to withstand impact, etc.	Low tendency to adhesion to avoid high sliding friction and wear; in practice it is often sufficient to avoid using identical materials
8. Bearing to be used in a sensitive force balance	Non magnetic; should not absorb moisture or be subject to any other weight variation. (Agate, for example, is unsatisfactory in the latter respect since it is hygroscopic)
9. Bearing to be used in a potentially corrosive environment (includes 'normal' atmospheres)	Good corrosion resistance, especially if requirement 8 has to be met

Table 25.5 Relevant properties of some knife edge and pivot materials

Material	Hardness H, Knoop	Modulus of elasticity, E ($N/m^2 \times 10^{10}$)	($lbf/in^2 \times 10^6$)	Load capacity factor, H^2/E (arbitrary units)	Ductility	Approximate maximum continuous operating temperature in air °C	°F	Corrosion resistance[1]
High carbon steel	to 690	21	30	2.3	Some	250	500	Poor
Tool steels	to 850	21	30	3.4	Some	650	1200	Poor–Good
Stainless steel (440C)	660	21	30	2.1	Some	430	800	Moderate
Agate	730	7.2	10.4	7.4	None	575[2]	1070[2]	Excellent
Synthetic corundum (Al_2O_3)	2100	38	55	11.6	None	1500	2700	Excellent
Boron carbide	2800	45	65	17.4	None	540	1000	Excellent
Silicon carbide	2600	41	60	16.4	None	800	1470	Excellent
Hot pressed silicon nitride	2000	31	45	13	None	1300	2400	Excellent

Notes: 1. Materials with poor corrosion resistance can often be protected by grease, oil bath or surface treatments (such as chromising of steels).
2. Phase change temperature.

Electromagnetic bearings

Electromagnetic bearings use powerful electromagnets to control the position of a steel shaft. Sensors are used to detect the shaft position and their output is used to control the currents in the electromagnets in order to hold the shaft in a fixed position. Steady and variable loads can be supported, and since no liquid lubricant is involved, new machine design arrangements become possible.

RADIAL BEARING CONFIGURATION

Four electromagnets are arranged around the shaft to form the bearing. Each electromagnet is driven by an amplifier. In horizontal shaft applications, the magnet centrelines are orientated at 45° to the perpendicular such that forces due to gravity are acted on by the upper two adjoining magnets. This adds to the load capability and increases the stability of the system.

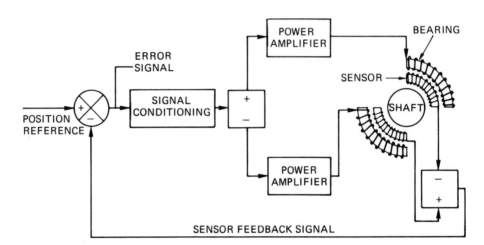

Fig. 26.1. Two of the four magnets of a radial bearing with their associated control system

Opposite electromagnets are adjusted to pull against one another in the absence of any externally applied force (the bias force). When an externally applied force causes a change in position of the shaft it is sensed by position transducers which, via the electronic control system, cause an increase in one current and a decrease in the other current flowing through the respective electromagnets. This produces a differential force to return the shaft to its original position. The signals from the position transducers continuously update this differential force to produce a stable system.

Typical radial bearing applications

A main field of application is on high speed rotating machines such as compressors, turbo-expanders, pumps and gas turbines.

Bearing bore sizes	40 – 1500 mm
Radial clearance gaps	0.1 – 5.0 mm
Speeds	400 – 120,000 RPM
Temperatures	185° C to 480° C
Load	up to 80 kN

AXIAL BEARING CONFIGURATION

A flat, solid ferromagnetic disc, secured to the shaft is used as the collar for the axial thrust bearing. Solid disc electromagnets are situated either side of the collar and operate in a similar manner to those in a radial bearing but in one dimension only.

POSITION TRANSDUCERS

Two dimensions are controlled at each radial bearing location and one dimension is controlled at the axial bearing. One transducer could be used for each dimension if it were totally linear and free from drift due to ageing or temperature effects. Two transducers per dimension are, however, used in practice because they require only that a balance or difference be maintained, thus cancelling unwanted offsets. A passive bridge system such as this greatly increases accuracy and reliability without undue increase in cost or complication.

PERFORMANCE RELATIVE TO HYDRODYNAMIC BEARINGS

Requirement	Magnetic bearings	Hydrodynamic bearings
1. High loads	Load capacity low, but bearing area could be higher than with conventional bearings (see 3 below)	High load capacity (except at low speeds)
2. High speeds	Limited mainly by bursting speed of shaft; system response to disturbance must be considered carefully	Shear losses can become significant
3. Sealing	No lubricant to seal, and the bearing can usually operate in the process fluid	Seals may need to be provided
4. Unbalance response	Shaft can be made to rotate about its inertial centre, so no dynamic load transmitted to the frame	Synchronous vibration results from unbalance
5. Dynamic loads	Damping can be tuned, but adequate response at high frequency may not be possible	Damping due to squeeze effects is high, and virtually instantaneous in its effect
6. Losses	Very low rotational losses at shaft, and low power consumption in magnets/electronics	Hydrodynamic and pumping losses can be significant, particularly at high speeds
7. Condition monitoring	Rotor position and bearing loads may be obtained from the control system	Vibration and temperature instrumentation can be added
8. Reliability and maintainability	Magnets and transducers do not contact the shaft so operating damage is unlikely; electronics may be sited in any convenient position	Very reliable with low maintenance requirements

Particular features of electromagnetic bearings

No mechanical contact.
No oil contamination of process fluid.
Shaft position does not change with speed.
Wide speed range including high speeds.
Can accept wide range of temperature.
Can provide a machine diagnostic output.
Requires a very reliable power supply and/or emergency support bearings.
Can produce electromagnetic radiation interference.
Requires space for its control system.

27   Bearing surface treatments and coatings

Table 27.1 The need for surface treatments and coatings

Type of application	Example	Function of coating or treatment
General use on many components		Allows components to be designed with a more optimum balance between the bulk and surface properties of the material
Lubricated plain bearing systems using high strength harder bearing materials	Crankshafts in heavy duty engines	Allows the shaft surface hardness to be increased to about five times the hardness of the bearing material, which is required for good compatibility
Lubricated components with small areas of contact with hard surfaces on both components in order to carry the contact pressures	Spur gears, cams and followers	Allows operation with low elastohydrodynamic film thickness with a reduced risk of scuffing and wear
Components operating at high loads and low speed or oscillating motion, and with only occasional lubrication	Bearings in mechanical linkages and roller chains, etc.	Aids oil retention on the surfaces and reduces the risk of seizure and wear
Surfaces which have intermittent rubbing contact with a reciprocating component	Cylinder liners in I.C. engines and actuators	Aids oil retention on the surfaces and reduces the risk of scuffing
Surfaces of components that are subject to fretting movements in contact with others	Connecting rod big end housing bores in high speed engines	Provides a surface layer that can allow small rubbing movements without the build up of surface damage
Components in contact with moving fluids containing abrasive material	Rotors and casings of pumps and fans	Gives abrasive wear resistance to selected areas of the surfaces of larger components
Components handling abrasive solid materials	Earth-moving machines. Coal and ore mills	Provides a hard abrasion resistant surface on a tough base material
Cutting tools	Drills and milling cutters	Provides a hard surface layer resistant to adhesion and abrasion and by providing a hard outer skin helps to retain sharpness
High temperature components with relative movements	Furnace conveyors. Boiler or reactor internals	Provides a surface resistant to adhesion and wear in the absence of conventional lubricants

Table 27.2 Types of surface treatments and coatings

General type	Materials which can be coated or treated	Examples	
Modifying the material at the surface without altering its chemical constituents	Ferrous materials, steels and cast irons	Induction hardening Flame hardening Laser hardening Shot peening	See Table 27.3
Adding new material to the surface, to change its chemical composition and properties	Ferrous materials	Carburising Nitriding Boronising Chromising	See Table 27.4
	Aluminium alloys	Anodising	
Placing a layer of new material on the surface	Ferrous materials Non–ferrous materials Plastics	Electroplating Physical vapour deposition Chemical vapour deposition Plasma spraying Flame spraying Vacuum deposition	See Table 27.5

Table 27.3 Surface treatment of ferrous materials

Process	Mechanism	Hardness and design depth	Aspects
Induction hardening of medium carbon alloy steels	Rapid heating followed by quenching produces martensite in the surface	Up to 600 Hv at depths up to 5 mm	Provides a hard layer deep enough to contain the high sub-surface shear stresses which occur in concentrated load contacts. Very large components can be treated
Flame hardening of medium carbon alloy steels	Surface heating followed by a quenching process produces martensite in the surface	Up to 500 Hv at depths up to 3 mm	Provides a hard wear-resistant surface layer. Particularly suitable for components with rotational symmetry that can be spun
Laser hardening of medium carbon steels and cast irons	A scanned laser beam focused on the surface. The very local heated area is self quenched by its surroundings	Up to 800 Hv at depths up to 0.75 mm	Suitable for local hardening of small areas of special components
Shot peening of ferrous materials	Work hardens the surface and leaves it in compression	Up to 0.5 mm approx	Gives increased resistance to fatigue and stress corrosion

27 Bearing surface treatments and coatings

Table 27.4 Diffusion of materials into surfaces

Process	Mechanism	Surface effects	Design aspects
Carburising of steels	Carbon is diffused into the surface at temperatures around 900°C followed by quenching and tempering	Hard surfaces at depths up to several mm depending on the diffusion time. Dimensional changes of ± 0.1% may occur	Suitable, for example, for gears to increase surface fatigue strength or pump parts to increase abrasive wear resistance
Carbonitriding of steels	Similar to carburising but with addition of nitrogen as well as carbon	Slightly harder surfaces are obtainable	Can be oil quenched with some reduction in distortion of the component
Boronising of steels	Diffusion of boron into the surface to form iron boride	Surface hardnesses up to 1200 Hv	The core material needs to be relatively hard to support the surface layer
Vanadium, niobium or chromium diffusion into steels	Salt bath treatment at 1020°C to produce thin surface layers of the metallic carbide	Layer hardness up to 3500 Hv at up to 15 μm thickness	Used to reduce wear of the surfaces of metal forming tools
Chromising, aluminising and siliconising of steels	High temperature pack processes for diffusing these materials into the surface	Increased surface hardness	Improves high temperature corrosion resistance and reduces fretting damage
Nitriding of steels which contain chromium or aluminium	Treatment at about 550°C in cracked ammonia gas or cyanide salt bath to produce chromium or aluminium nitrides in the surface layers	Hardness up to 850 Hv with surface layers up to 0.3 mm deep	Forms a brittle white surface layer which needs to be removed before use in tribological applications
Nitrocarburising of steels including plain carbon steels (tufftriding)	Salt bath, gas or plasma treatments are available	Hard surface layer about 20 μm thick with hardness of about 700 Hv	Reduces fretting damage. The surface layer can be oxidised and impregnated with lubricant to produce a low friction corrosion resistant surface
Ion implantation of steel	Surface bombardment by a high energy nitrogen ion beam at 150°C produces up to 30% implanted material at the surface	Surface effects to a depth of 0.1 μm	Can improve resistance to abrasion and fatigue
Sulfinuz treatment of steel	Salt bath treatment at about 600°C to add carbon, nitrogen and sulphur	Surface layer contains sulphides which act as a solid lubricant in the surface	Can give improved scuffing resistance with little risk of component distortion during treatment
Phosphating of steel	Phosphate layer produced by chemical or electrochemical action	Surface layer which is porous and helps to retain lubricant	Assists running in and reduces risk of scuffing
Electrolytic deposition and diffusion of material into non-ferrous metals	Copper, aluminium and titanium can be treated at temperatures in the range 200–400°C	Creates surface layers at selected positions	Gives improved scuffing and wear resistance
Hard anodising of aluminium	The component is made the anode for electrolytic treatment in sulphuric acid. The oxygen generated at the surface produces a hard porous oxide layer	The hard layer is usually 25–75 μm thick	The coating is porous and can be overlaid with PTFE to to give a low friction surface

132

Table 27.5 The coating of surfaces

Process	Mechanism	Surface effects	Design aspects
Electroplating with hard chromium	Plating is carried out at less than 100°C so there are no distortion problems. Current density can be adjusted to produce tensile stresses which, with diamond honing produce regular distributed cracks for good oil retention	Surface hardness of up to 900 Hv and thicknesses of up to 1 mm. Also used in thin, as deposited layers about 50 μm thick	A good surface for cylinder liners but needs to be cleaned up by Si C honing. A good piston ring coating for use with cast iron liners
Electroplating with tin, lead or silver	Can be plated on to any metals	Soft surface layers with thicknesses typically up to 50 μm	Gives good resistance to fretting and improves the bedding in and running properties of harder bearing materials
Electroless nickel plating	A low temperature process with good throwing power to follow component shapes. Can also be used with hard wear-resistant particles or PTFE particles dispersed in the coating	The hardness is about 550 Hv as deposited but can be increased to 1000 Hv by heat treatment	Good for cylinder liners and all component surfaces requiring improved wear resistance. Inclusion of PTFE gives lower friction
Physical vapour deposition of coatings such as titanium nitride	The coating is transferred to the component via a glow discharge in a vacuum chamber. Deposition temperature is 250–450°C	Ti.N coatings have a hardness of about 3000 Hv and are usually 2–4 μm thick	Ideal for cutting tools and for components subject to adhesive wear or low stress abrasion such as hard particles sliding over the surface. It needs an undercoat of electroless nickel if full corrosion resistance is required
Chemical vapour deposition of metals and ceramics	The coating is produced by decomposition of a reactive gas at the component surface. The operation occurs within a reaction chamber at 800°C minimum temperature	Coatings of Ti.N TiC. Al$_2$0$_3$. CrN WC and CrC in thicknesses of the order of 5–10 μm	Very hard wear resistant coatings can be created by this process
Spray coating using a gas flame or electric arc	The coating material is supplied to the process as a powder or as a wire or wires. The powder coating may then be subsequently fused to the surface	Coatings can be of hardnesses up to 900 Hv in thicknesses in the range of 0.05–1.0 mm	Suitable for covering large areas of component surfaces
Plasma spray coating	Uses an ionised inert gas to produce very high temperatures which enable oxides and other ceramics to be coated onto metals. Component heating is minimal	Typical coating thickness 0.1 mm	Particularly suitable for improving the abrasive wear resistance of components
Plasma arc spraying	Carried out in a partially evacuated chamber or under an inert gas shield. The component gets hot and good bonding is achieved. Higher temperatures can be achieved by having an electric current flow between the arc and the surface. The performance of the materials can be further improved by hot isostatic pressing at temperatures of the order of 1000°C and above	Up to 10 mm thicknesses of very hard ceramic materials can be deposited	Suitable for high temperature gas turbine components to give improved oxidation and fretting resistance. Also abrasion resistant components for mining and agricultural machinery

Table 27.5 The coating of surfaces (continued)

Process	Mechanism	Surface effects	Design aspects
High velocity explosive coating	The coating spray is driven by successive explosive hydrocarbon combustion cycles. This gives coatings with very low porosity and good adhesion	Coatings of CrC, WC and cobalt-based cermets about 0.1 μm thick	Produces very tough abrasion resistant coatings
Thermochemically formed coatings	Ceramic coatings sprayed on and then heat treated at temperatures of about 500°C	Coatings such as chromium oxide of 1100Hv about 0.1 mm thick	Produces a corrosion and abrasion resistant coating
Laser cladding and alloying	The application of coatings derived from powders in small controlled areas	About 0.1 mm thick as a maximum	Good local corrosion and abrasion resistance
Weld coatings	Layers of hard material applied from welding rods and wires by gas or electric welding	Thick layers up to 30 mm thick can be applied selectively	A relatively adaptable manual process suitable for the repair and improved wear resistance of heavy duty components subject to abrasive wear
Cladding	Hard surface materials attached to components by welding, adhesives, or roll bonding	Thick wear-resistant materials can be attached to surfaces	Components, chutes, etc., requiring abrasion resistance

INDEX